Soil Fertility Management
in Agroecosystems

American Society of Agronomy, Inc.
Crop Science Society of America, Inc.
Soil Science Society of America, Inc.
5585 Guilford Road, Madison, WI 53711-5801 USA

agronomy.org | soils.org | crops.org
dl.sciencesocieties.org
SocietyStore.org

ISBN: 978-0-89118-353-2 (print)
ISBN: 978-0-89118-354-9 (digital)
doi:10.2134/soilfertility
Library of Congress Control Number: 2016961005

ACSESS Publications
ISSN: 2165-9842 (online)
ISSN: 2165-9834 (print)

Cover design: Patricia Scullion

Soil Fertility Management in Agroecosystems

Amitava Chatterjee and David Clay, Editors

Contents

Preface

Soils sustain life and provide the nutrients, water, and oxygen crucial for crop production. With time, the concept of soil fertility has been expanded from making a fertilizer recommendation to industry efforts to use climate-driven simulation models for optimizing soil health and fertilizer efficiency, while minimizing agriculture's impact on the environment. Increasingly, we are adopting the principles of precision agriculture for improving crop nutrient use efficiency and minimizing nutrient losses. Scientists are also integrating on-farm testing into their research programs. In these studies, natural abundance stable isotopic techniques are being used to improve our understanding carbon, nitrogen, and water cycling.

For the requirements of our future workforce, it is imperative that we modify our curricula to account for a dynamic environment—agronomists and soil scientists are increasingly challenged by extreme climatic conditions, and farmers are experimenting with integrating cover crops into their rotations and reducing tillage intensity and the amount of chemical fertilizers applied to their soils. Farmers wonder about the value of soil health, industry agronomists are using simulation models to develop locally based recommendations, and environmentalists are concerned about carbon footprints. Many advanced soil fertility textbooks do not address these critical issues.

The purpose of this book was not to reiterate the basic soil fertility facts discussed elsewhere, but to present and discuss examples of how modern research tools can help develop better answers to production and environmental questions. The editors are heavily indebted to the contributing authors, and we invite researchers, students, and extension personnel across the country to provide feedback, comments, and suggestions on the material, so that we can incorporate their comments in future editions.

Amitava Chatterjee and David Clay, Editors

Contributors

Bruggeman, Stephanie A.H. — Plant Science Dep., South Dakota State Univ., Berg Agricultural Hall, P.O. Box 2207A, Brookings, SD 57007 (Stephanie.Bruggeman@sdstate.edu)

Carlson, Gregg — Plant Science Dep., South Dakota State Univ., Berg Agricultural Hall, P.O. Box 2207A, Brookings, SD 57007 (gregg.carlson@sdstate.edu)

Chang, Jiyul — Plant Science Dep., South Dakota State Univ., Berg Agricultural Hall, P.O. Box 2207A, Brookings, SD 57007 (Jiyul.Chang@sdstate.edu)

Chatterjee, Amitava — Dep. of Soil Science, Walster 133, 1402 Albrecht Blvd., North Dakota State Univ., Fargo, ND 58108 (amitava.chatterjee@ndsu.edu)

Clay, David E. — Plant Science Dep., South Dakota State Univ., Berg Agricultural Hall, P.O. Box 2207A, Brookings, SD 57007 (david.clay@sdstate.edu)

Clay, Sharon A. — Plant Science Dep., South Dakota State Univ., Berg Agricultural Hall, P.O. Box 2207A, Brookings, SD 57007 (sharon.clay@sdstate.edu)

Culman, Steven W. — The Ohio State University, The Ohio Agricultural Research and Development Center, 1680 Madison, Ave., Wooster, OH 44691 (culman.2@osu.edu)

Dick, Warren A. — The Ohio State University, The Ohio Agricultural Research and Development Center, 1680 Madison, Ave., Wooster, OH 44691 (dick.5@osu.edu)

Franzen, David — North Dakota State Univ., Dep. 7180, P.O. Box 6050, Fargo, ND 58108 (david.franzen@ndsu.edu)

Hatfield, Jerry L. — National Laboratory for Agriculture and the Environment, 2110 University Blvd., Ames, IA 50011 (jerry.hatfield@ars.usda.gov)

Kaizzi, Kayuki — National Agricultural Research Laboratory, P.O. Box 7065, Kampala, Uganda (kckaizzi@gmail.com)

Kharel, Tulsi — Plant Science Dep., South Dakota State Univ., Berg Agricultural Hall, P.O. Box 2207A, Brookings, SD 57007 (Tulsi.Kharel@sdstate.edu)

Moriles-Miller, Janet — Plant Science Dep., South Dakota State Univ., Berg Agricultural Hall, P.O. Box 2207A, Brookings, SD 57007 (janet.miller@sdstate.edu)

Reese, Cheryl — Plant Science Dep., South Dakota State Univ., Berg Agricultural Hall, P.O. Box 2207A, Brookings, SD 57007 (Cheryl.Reese@sdstate.edu)

Reicks, Graig — Plant Science Dep., South Dakota State Univ., Berg Agricultural Hall, P.O. Box 2207A, Brookings, SD 57007 (graig.reicks@sdstate.edu)

Wortmann, Charles — Dep. of Agronomy and Horticulture, Univ. Nebraska–Lincoln, 279 Plant Science, Lincoln, NE 68583-0915 (cwortmann2@unl.edu)

- Considerations in Soil Fertility Management
- Nutrient Use Efficiency and Nutrient Balance

Soil Fertility: Current US Situation and Challenges

Amitava Chatterjee

Maintaining soil fertility is a critical component of the crop production system to ensure economic profitability and sustaining productivity. The ongoing plummet in crop prices, shrinking cropland area, increasing fertilizer prices, and environmental concerns associated with fertilizer application are triggering a rethinking of our current principles of crop nutrition and fertilizer management practices. The recent development of the 4-R stewardship program was designed to reduce the adverse impacts of agriculture on the environment. In future systems, soil scientists and agronomists will be asked to consider both goods and services, as well as increased crop diversity. Meeting projected food requirements for 2050 requires the discovery of the connections among soil health, crop productivity, tillage intensity, climatic conditions, and the agronomic requirements for production systems. In this chapter, the current US situations of crop production area, nutrient demand, and nutrient removal are discussed as a foundation to understand the scope for the development of fertilizer management.

Soil fertility is the capacity of soil to supply nutrients to a crop. However, inherent soil nutrient stocks are not sufficient to produce harvestable yield or economic profit. Historically in the United States, soil fertility specialists investigated the relationship between soil nutrient levels and yields. This research resulted in a wide variety of soil chemical extracts that measured soil test levels, which were then corrected to yield responses and fertilizer recommendations. Prior to the introduction of chemical fertilizers, application of organic manures was the main form of applied nutrients. Since the discovery of chemical fertilizers, crop yield has significantly increased. Chemical fertilizers contribute at least 50% of crop yield and a significant portion in production cost and economic profitability in farming (Stewart et al., 2005). Besides, application of chemical fertilizers maintains soil nutrient concentrations and stabilize soil health. A recent study of long-term fertilizer experiment around the world determined that the use of nitrogenous fertilizer increased soil microbial population by 15% and soil organic carbon by 13% (Geisseler and Scow, 2014; Clay et al., 2012).

Increasing demand for food and shrinking agricultural resources—cropland and water—have forced us to increase crop productivity. Introduction of high-yielding cultivars has significantly increased the yield potential over the last four decades. Adequate supply of nutrients is needed to exploit the full

Amitava Chatterjee, Dep. of Soil Science, Walster 133, 1402 Albrecht Blvd., North Dakota State Univ., Fargo, ND 58108 (amitava.chatterjee@ndsu.edu).

doi:10.2134/soilfertility.2014.0002

genetic potential. Furthermore, expansion and intensification of agricultural land will reduce the area and sustainability of other ecosystems (forests, wetlands, and marine) and their associated biodiversity (Vitousek et al., 2009; Sayer and Cassman, 2013). World population will reach 9 billion in 2050, resulting in an increase in food demand of 58 to 95%, depending on changes in food consumption habits (McKenzie and Williams, 2015). For this reason, it is imperative that we explore sustainable soil fertility management practices to increase crop nutrient use efficiency. In this chapter, the current situation and the challenges of different crop production and soil fertility issues are discussed to facilitate understanding of the following chapters.

Considerations in Soil Fertility Management

Cropland Area

Global increase in cropland area was less rapid (+25.7%) than the rise in population (+74%) and per capita income (+123%) between 1970 and 2005 (Rudel et al., 2009). Cropland includes five segments: cropland harvested, crop failure, cultivated summer fallow, pasture, and idle cropland. The US cropland has remained relatively constant since World War II, but has slowly declined since 1995, decreasing from 178 to 165 Mha (–7.56%) in 4 yr (Table 1). The Southeast and Southern Plains regions lost almost

16% of cropland, followed by the Delta and Appalachian areas, with 13% losses. Cropland loss has been mainly due to exurban growth throughout the United States and conversion and abandonment of agricultural lands, especially in the eastern United States (Lubowski et al., 2006; Brown et al., 2005). However, the most important farm regions—Corn Belt, Lake States, and Northern Plains—showed only 3 to 4% cropland loss.

The percent of total US cropland in the Corn Belt, Lake States, and Northern Plains has increased with time. Cropland use has changed during 2000 through 2013 (Fig. 1). An increase in corn (*Zea mays* L.) area was driven by increased demand for bioethanol, with shifts to more monoculture or continuous corn rotation (Plourde et al., 2013). Soybean [*Glycine max* (L.) Merr.] and wheat (*Triticum aestivum* L.) areas were consistent. US corn production has increased since 2000, but wheat and soybean production have not changed much. National average corn yield varied between 7.7 and 10.3 Mg ha^{-1}. Since 2003, annual corn yield has been above 9 Mg ha^{-1}, with the exception of 2012. Wheat and soybean average yield has hardly changed. Climatic variations and lack of rainfall resulted in the lowest yield of all three crops in 2002 and corn in 2012.

Chemical Fertilizer Use

Chemical fertilizer use is important to maintain sustainable crop production. Stewart et al. (2005) determined that nitrogen fertilizer was responsible for 41, 37, and 16% of corn, cotton (*Gossypium hirsutum* L.), and wheat yield, respectively. Cost of chemical fertilizers is a significant portion of total production cost (Table 2). An approximate budget of irrigated corn, soybean, and spring wheat production in eastern North Dakota revealed that application of chemical fertilizers accounted for 26, 25, and 7% of total costs for corn, spring wheat, and soybean production, respectively (Table 2). It is clear that increases in fertilizer prices will significantly reduce the return to growers if crop market prices do not increase proportionately. Corn, wheat, and soybean annual prices have increased with time, but fertilizer prices have also increased (Fig. 2 and 3). High crop prices resulted in large net cash returns and encouraged growers to expand crop acres and application of higher fertilizer rates to increase productivity. This increased demand for fertilizers resulted in higher fertilizer prices (Huang, 2009), generally associated with higher costs for

Table 1. Changes in cropland area across US farm regions (source: USDA-NASS).

	2002	2007	Change
	— Million ha —		%
Northeast	5.55	5.25	–5.35
Lake States	17.1	16.4	–3.66
Corn Belt	38.8	36.9	–4.92
Northern Plains	41.3	39.6	–4.21
Appalachian	10.5	9.18	–12.91
Southeast	6.00	5.06	–15.79
Delta States	8.52	7.38	–13.38
Southern Plains	22.5	19.0	–15.65
Mountains	18.7	17.5	–6.53
Pacific	9.70	8.96	–7.68
48 States	179	165	–7.56

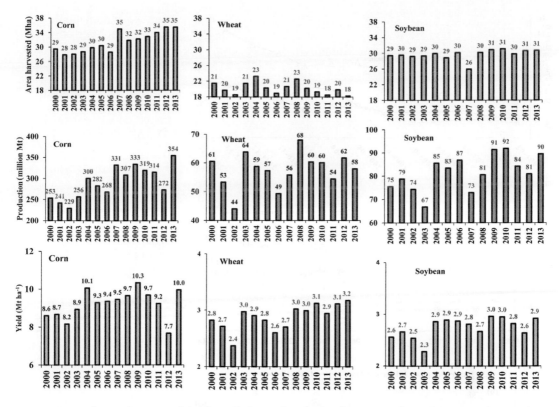

Fig. 1. Annual area harvested (million ha), production (million t), and productivity (t ha⁻¹) of corn, wheat, and soybean during the 2000–2013 growing season in the United States. Source: USDA-NASS (https://www.nass.usda.gov/Data_and_Statistics/index.php).

Table 2. Approximate budget for irrigated corn, soybean, and spring wheat in eastern North Dakota. Percentage to total cost of selected items in bracket (source: www.ag.ndsu.edu, accessed 1 Sept. 2016).

	Corn grain	Soybean	Spring wheat
Market yield, Mg ha⁻¹	12.5	3.76	4.70
Market price, $ Mg⁻¹	63.78	171.42	104.18
Market income, $ ha⁻¹	1976	1593	1210
Direct costs, $ ha⁻¹			
Seed	272 (15%)	191 (18%)	61.75 (5%)
Herbicides	53 (3%)	53.64	52.49
Fungicides	0	0	33.35
Insecticides	0	19	0
Fertilizer	450 (26%)	72 (7%)	305.17 (25%)
Crop insurance	122	45.13	52.39
Fuel and lubrication	59	41	39.86
Repairs	41	37	36.53
Irrigation power	48	48	47.94
Irrigation repairs	26	26.38	26.38
Drying	99	0	0
Miscellaneous	2.47	2.47	2.47
Operating interest	35	16.06	19.76
Total direct costs, $ ha⁻¹	1208	551.68	678
Indirect (Fixed costs)	558	535	535.79
Sum of all listed costs	1766	1086.68	1213.85
Return to labor and management, $ ha⁻¹	210	507.58	−3.85

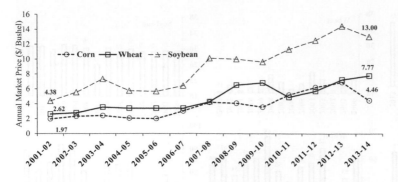

Fig. 2. Annual mean corn, wheat and soybean prices in the United States, 2001–2014. Source: http://www.ers.usda.gov/data-products/season-average-price-forecasts.aspx (accessed 1 Sept. 2016).

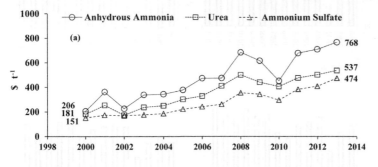

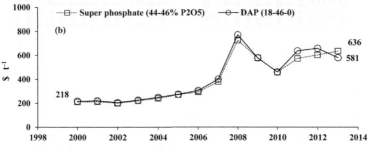

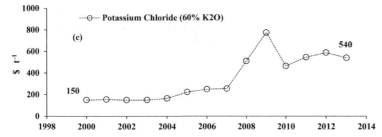

Fig. 3. Increase in cost for (a) nitrogenous, (b) phosphorus, and (c) potassium fertilizers since 2000. Source: Fertilizer Use and Price, USDA-ERS, http://www.ers.usda.gov/data-products/fertilizer-use-and-price.aspx (accessed 1 Sept. 2016).

energy, raw material, and transportation. Prices of common nitrogenous, phosphorus, and potassium fertilizers have increased almost three to four times since 2000 (Fig. 3).

Commercial fertilizer production is heavily dependent on fossil fuel energy, and an increase in the energy price can increase the cost of fertilizer. For nitrogen fertilizer production, ammonia is the primary feedstock and natural gas the main input. Maintenance of phosphorus and potassium nutrition of crops in the United States is mostly dependent on commercial fertilizer sources. Phosphorus mined from rock

is a finite resource, with highly concentrated deposits limited to Morocco with 85% of the global share, followed by China with 6% and the United States with 3% (McKenzie and Williams, 2015). High grade phosphate (P_2O_5) rock reserves are likely to be exhausted within 50 to 150 yr, and a decrease in the grade used will increase the contamination levels, requiring additional energy costs (Schröder et al., 2011). US potash (K_2O) production accounts for less than 16% of the domestic potash fertilizer supply, and most potash is imported from Canada (Huang, 2009). Increases in the K_2O demands of other countries, like China, India, and Brazil, have resulted in an extreme tightening of the global potash supply. US fertilizer consumption was dominated by nitrogen (11.6 million t), followed by K_2O (4.1 million t) and P_2O_5 (3.9 million t) in 2011 (Fig. 4a). Urea ammonium nitrate or nitrogen solutions are now the most used nitrogen fertilizers, followed by urea and anhydrous or liquid ammonia (Fig. 4b). Mono- and di-ammonium phosphates are the most common phosphorus fertilizer sources (Fig. 4c).

Nutrient Use Efficiency and Nutrient Balance

Increasing demand for fertilizers, rising fertilizers prices, and diminishing phosphorus reserves drive demand for more nutrient use efficiency by increasing the crop uptake and reducing nutrient loss. Moreover, the loss of nitrogen and phosphorus from agricultural systems has major environmental consequences. The basic philosophy of increasing nutrient use efficiency is known as the 4 Rs—right source, right application rate (enough to meet crop demand), right application timing (synchronized with plant nutrient uptake), and right placement (to maximize the crop's access to nutrients). Nutrient budgeting approaches are used to evaluate the nutrient balance or nutrient use efficiency by estimating input (fertilizer applied) and output (crop removal). Nitrogen, phosphorus, and potassium fertilizer balances of corn, wheat, and soybean for major agricultural states are presented in Tables 3 through 5. Average yields of different states are calculated using yield data from the Crop Production Annual Summary published by USDA-NASS (available yearly online:// usda.mannlib.cornell.edu, accessed 31 Aug. 2016). Nitrogen, phosphorus, and potassium fertilizer application rates for corn, wheat, and soybean of major producing states are available online from USDA-ERS (http://www.ers.usda.gov/data-products/fertilizer-use-and-price.aspx, accessed 31 Aug. 2016). Removal rates of nitrogen, phosphorus, and potassium were calculated using the conversion factors provided by Johnston and Usherwood

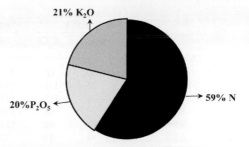

(a) US Percent consumption of plant nutrients

21% K_2O
20% P_2O_5
59% N

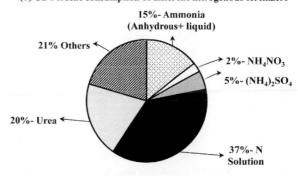

(b) US Percent consumption of different nitrogenous fertilizers

15%- Ammonia (Anhydrous+ liquid)
21% Others
2%- NH_4NO_3
5%- $(NH_4)_2SO_4$
20%- Urea
37%- N Solution

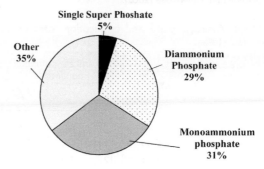

(c) US Percent consumption of different phosphorus fertilizers

Single Super Phosphate 5%
Other 35%
Diammonium Phosphate 29%
Monoammonium phosphate 31%

Fig. 4. Percent application in the United States of (a) nitrogen, phosphate (P_2O_5) and potash (K_2O), (b) nitrogenous, and (c) phosphate fertilizers in 2011. Source: Fertilizer Use and Price, USDA-ERS, available http://www.ers.usda.gov/data-products/fertilizer-use-and-price.aspx (accessed 1 Sept. 2016).

(2002). Soybean is a legume crop, and nitrogen was not included in the balance calculation.

Nutrient balance showed wide variations depending on crop and state. For corn, nitrogen balance is positive for most states, indicating additional nitrogen was supplied to compensate for losses through volatilization, denitrification, and leaching (Table 3). However, wheat nitrogen balance showed few negative

Table 3. Average yield and application rates of nitrogen, phosphate, and potash used on corn for selected states.

	Avg. yield[†]	Fertilizer nutrient applied[†‡]			Grain nutrient removal[§]			Nutrient balance = Applied – Removed		
		N	P_2O_5	K_2O	N	P_2O_5	K_2O	N	P_2O_5	K_2O
	Mg ha^{-1}	\-- kg ha^{-1} \--								
Colorado	8.30	142	31	17	111	65.2	43.0	31	−34.2	−26
Georgia	10.7	198	76	111	143	84.1	55.4	55	−8.1	55.6
Illinois	9.20	187	104	119	123	72.3	47.6	64	31.7	71.4
Indiana	8.82	199	77	133	118	69.3	45.7	81	7.7	87.3
Iowa	9.91	159	73	90	133	77.9	51.3	26	−4.9	38.7
Kansas	6.90	147	41	45	92.4	54.2	35.7	54.6	−13.2	9.3
Kentucky	7.88	184	111	119	106	61.9	40.8	78	49.1	78.2
Michigan	9.22	137	36	105	123	72.4	47.7	14	−36.4	57.3
Minnesota	10.1	140	56	71	135	79.4	52.3	5	−23.4	18.7
Missouri	6.79	141	71	64	90.9	53.4	35.2	50.1	17.6	28.8
Nebraska	9.87	157	46	29	132	77.6	51.1	25	−31.6	−22.1
New York	8.47	66	40	43	113	66.6	43.9	−47	−26.6	−0.9
North Carolina	7.17	143	45	87	96.0	56.3	37.1	47	−11.3	49.9
North Dakota	7.04	179	49	37	94.3	55.3	36.5	84.7	−6.3	0.5
Ohio	9.57	158	72	102	128	75.2	49.6	30	−3.2	52.4
Pennsylvania	8.15	97	53	58	109	64.0	42.2	−12	−11	15.8
South Dakota	7.76	144	57	32	104	61.0	40.2	40	−4	−8.2
Texas	7.55	144	40	19	101	59.3	39.1	43	−19.3	−20.1
Wisconsin	8.84	103	49	59	118	69.5	45.8	−15	−20.5	13.2
UA avg.	8.97	157	67	88	120	70.5	46.5	34.4	−7.8	26.3

† Average yield of 2011–2013 production year, Source: USDA, NASS, Annual Crop Production Summary (https://www.nass.usda.gov/Data_and_Statistics/index.php).
‡ Source: Fertilizer Use and Price, USDA-ERS, http://www.ers.usda.gov/data-products/fertilizer-use-and-price.aspx.
§ Johnston and Usherwood (2002).

Table 4. Average yield and application rates of nitrogen, phosphate, and potash used on wheat for selected states.

	Avg. yield[†]	Fertilizer nutrient applied[†‡]			Grain nutrient removal[§]			Nutrient balance = Applied – Removed		
		N	P_2O_5	K_2O	N	P_2O_5	K_2O	N	P_2O_5	K_2O
	Mg ha^{-1}	\-- kg ha^{-1} \--								
Colorado	2.28	39	26	20	57.0	19.0	13.3	−18	7	6.7
Idaho	5.45	134	32	26	136	45.4	31.8	−2	−13.4	−5.8
Illinois	4.28	121	90	113	107	35.7	25.0	14	54.3	88
Kansas	2.58	57	34	35	64.5	21.5	15.1	−7.5	12.5	19.9
Minnesota	3.58	129	46	26	89.5	29.8	20.9	39.5	16.2	5.1
Missouri	3.65	104	63	81	91.3	30.4	21.3	12.7	32.6	59.7
Montana	2.44	66	31	20	61.0	20.3	14.2	5	10.7	5.8
Nebraska	2.71	62	38	16	67.8	22.6	15.8	−5.8	15.4	0.2
North Dakota	2.68	99	47	27	67.0	22.3	15.6	32	24.7	11.4
Ohio	4.42	101	60	71	111	36.8	25.8	−10	23.2	45.2
Oklahoma	2.00	71	35	16	50.0	16.7	11.7	21	18.3	4.3
Oregon	4.56	90	24	43	114	38.0	26.6	−24	−14	16.4
South Dakota	2.81	86	38	24	70.3	23.4	16.4	15.7	14.6	7.6
Texas	1.95	62	35	19	48.8	16.3	11.4	13.2	18.7	7.6
Washington	4.61	91	21	22	115	38.4	26.9	−24	−17.4	−4.9
US avg.	3.08	77	38	27	77.0	25.7	18.0	4.12	13.56	17.81

† Average yield of 2011–2013 production year, Source: USDA, NASS, Annual Crop Production Summary (https://www.nass.usda.gov/Data_and_Statistics/index.php).
‡ Source: Fertilizer Use and Price, USDA-ERS, http://www.ers.usda.gov/data-products/fertilizer-use-and-price.aspx.
§ Johnston and Usherwood (2002).

Table 5. Average yield and application rates of nitrogen, phosphate, and potash used on soybean for selected states.

	Avg. yield	Fertilizer nutrient applied†‡		Grain nutrient removal‡§		Nutrient balance = Applied – Removed	
		P_2O_5	K_2O	P_2O_5	K_2O	P_2O_5	K_2O
	Mg ha^{-1}	kg ha^{-1}					
Arkansas	2.80	60	99	37	65.3	23	34
Illinois	3.13	78	122	42	73.0	36	49
Indiana	3.15	60	104	42	73.5	18	31
Iowa	3.15	62	88	42	73.5	20	15
Kansas	1.92	39	48	26	44.8	13	3
Kentucky	2.88	59	91	38	67.2	21	24
Louisiana	2.91	46	73	39	67.9	7	5
Michigan	2.94	40	87	39	68.6	1	18
Minnesota	2.77	43	60	37	64.6	6	−5
Mississippi	2.89	52	96	39	67.4	13	29
Missouri	2.29	59	85	31	53.4	28	32
Nebraska	3.33	56	40	44	77.7	12	−38
North Carolina	2.31	50	93	31	53.9	19	39
North Dakota	2.10	41	21	28	49.0	13	−28
Ohio	3.18	58	112	42	74.2	16	38
South Dakota	2.41	53	37	32	56.2	21	−19
Tennessee	2.60	59	91	35	60.7	24	30
Virginia	2.69	50	93	36	62.8	14	30
Wisconsin	2.82	39	101	38	65.8	1	35
US avg.	2.80	55	90	37	65.3	18	25

† Average yield of 2011–2013 production year, Source: USDA, NASS, Annual Crop Production Summary (https://www.nass.usda.gov/Data_and_Statistics/index.php).

‡ Source: Fertilizer Use and Price, USDA-ERS, http://www.ers.usda.gov/data-products/fertilizer-use-and-price.aspx.

§ Johnston and Usherwood (2002).

values, indicating the nitrogen supplied was less than crop removal (Table 4). Corn showed a negative phosphorus balance in most states, but phosphorus balance was positive for wheat and soybean for most states. Although potassium showed a positive balance for most states in all three crops, Nebraska recorded a negative balance for corn and soybean. For the economic response of the crop, production relied to some extent on soil supply of phosphorus and potassium, and phosphorus and potassium fertilizer applications were avoided. Fixen et al. (2010) reported a consistent decline in soil phosphorus levels across the Corn Belt and Central Great Plains, and many states east of the Mississippi Rivers were at or below agronomic critical levels, indicating that 50% or more of these areas require regular potassium application to avoid yield losses. Increase in continuous corn and corn–soybean rotation will continue to drive the need for soil phosphorus and potassium application in this region. Phosphorus and potassium fertilizer recommendations follow "sufficiency–deficiency correction," "buildup–maintenance," or "replenishment

of crop removal" concepts, and variations in critical soil test levels for these nutrients vary among states and soil types (Hergert et al., 1997; Dobermann and Cassman, 2002). Sustainable management of soil nutrients will require fertilizer recommendation algorithms considering economic (grain and fertilizer price), crop and soil management factors (rotation, tillage, soil properties), and climatic (rainfall, temperature, growing degree days) variables.

The world is challenged with two seemingly contradictory goals, increasing food production and maintaining critical ecosystem services. Globally, the increasing need for the food production through increased specialization (Erenstein and Thorpe, 2009) may not be environment friendly (Matson et al., 1997) or resilient. In many systems, focusing on production results in a decrease in services. Syswerda and Robertson (2014) reported that soil C was negatively correlated with nitrate leaching, and soil water content was positively correlated with grain yield and negatively correlated with soil C and plant diversity. Diversity may also impact nutrient efficiency and the actual response function between the soil test value and yield response. Many of current soil test recommendations were developed between the 1950s and 1980. During this period, fields were routinely plowed, disked, and cultivated to control weeds. These practices discouraged soil microbial functions. Since then we have learned that organisms can assist in nutrient and water use and help manage surface residues. In addition, diversity by itself can change the nutrient response functions. Wortmann et al. (2009) reported that by switching from continuous corn to corn–soybean rotation reduced the critical P level from 20 to 10 mg P kg^{-1}. In summary, increasing food supplies are needed to feed an increasing global population. This will require increased resource use efficiency to ensure sustainable development. In the future, soil fertility specialists will be asked to assess the impact of soil fertility programs on the services provided by the fertilizer. Complete analysis of the system will lead to an improved conceptual framework on how to balance goods and services.

References

Brown, D.G., K.M. Johnson, T.R. Loveland, and D.M. Theobald. 2005. Rural land-use trends in the conterminous United States, 1950–2000. Ecol. Appl. 15(6):1851–1863. doi:10.1890/03-5220

Clay, D.E., J. Chang, S.A. Clay, J. Stone, R. Gelderman, C.G. Carlson, K. Retisma, M. Jones, L. Janssen, and T. Schumacher. 2012. Corn yields and no-tillage affects carbon sequestration and carbon footprints. Agron. J. 104:763–770.

Dobermann, A., and K.G. Cassman. 2002. Plant nutrient management for enhanced productivity in intensive grain production systems of the United States and Asia. Plant Soil 247:153–175. doi:10.1023/A:1021197525875

Erenstein, O., and W. Thorpe. 2009. Crop-livestock interactions along agro-ecological gradients: a meso-level analysis in the Indo-Gangetic Plains, India. Environ. Develop. Sustainability doi:10.1007/s10668-009-9218-z

Fixen, P.E., T.W. Bruulsema, T.L. Jensen, R. Mikkelsen, T.S. Murrell, S.B. Phillips, Q. Rund, and W.M. Stewart. 2010. The fertility of North America soils. Better Crops 94(4):6–8.

Geisseler, D., and K.M. Scow. 2014. Does long-term use of mineral fertilizers affect the soil microbial biomass? Better Crops Plant Food 98(4):13–15.

Hergert, G.W., W.L. Pan, D.R. Huggins, J.H. Grove, and T.R. Peck. 1997. Adequacy of current fertilizer recommendations for site-specific management In: F.J. Pierce and E.J. Sadler, editors, The state of site-specific management for agriculture. ASA, CSSA, SSSA, Madison, WI. p. 283–300. doi:10.2134/1997. stateofsitespecific.c13

Huang, W. 2009. Factors contributing to the recent increase in U.S. fertilizer prices, 2002-08. USDA-ERS Outlook Rep. AR-33. ers.usda.gov/publications/ar-agricultural-resources-situation-and-outlook/ar-33.aspx (accessed 16 July 2015).

Johnston, A.M., and N.R. Usherwood. 2002. Crop nutritional needs. In: Plant nutrient use in North American agriculture. PPI/PPIC/FAR Tech. Bull. 2002-1. Potash and Phosphate Inst., Norcross, GA. p. 13–22.

Lubowski, R.N., M. Vesterby, S. Bucholtz, A. Baez, and M. Roberts. 2006. Major uses of land in the United States. USDA-ERS Economic Information Bull. 14. http://www.ers.usda.gov/media/250091/eib14_1_.pdf (accessed 31 Aug. 2016).

Matson, P.A., W.J. Parton, A.G. Power, and M.J. Swift. 1997. Agricultural intensification and ecosystem properties. Science 277:504–509.

McKenzie, F.C., and J. Williams. 2015. Sustainable food production: Constraints, challenges and choices by 2050. Food Secur. 7:221–233. doi:10.1007/s12571-015-0441-1

Plourde, J.D., B.C. Pijanowski, and B.K. Pekin. 2013. Evidence for increased monoculture cropping in the Central United States. Agric. Ecosyst. Environ. 165:50–59. doi:10.1016/j.agee.2012.11.011

Rudel, T.K., L. Schneider, M. Uriarte, B.L. Turner II, R. DeFries, D. Lawrence, J. Geoghegan, S. Hecht, A. Ickowitz, E.F. Lambin, T. Birkenholtz, S. Baptista, and R. Grau. 2009. Agricultural intensification and changes in cultivated areas, 1970–2005. Proc. Natl. Acad. Sci. USA 106(49):20675–20680. doi:10.1073/pnas.0812540106

Sayer, J., and K.G. Cassman. 2013. Agricultural innovation to protect the environment. Proc. Natl. Acad. Sci. USA 110(21):8345–8348. doi:10.1073/pnas.1208054110

Schröder, J.J., A.L. Smit, D. Cordell, and A. Rosemarin. 2011. Improved phosphorus use efficiency in agriculture: A key requirement for its sustainable use. Chemosphere 84:822–831. doi:10.1016/j.chemosphere.2011.01.065

Stewart, W.M., D.W. Dibb, A.E. Johnson, and T.J. Smyth. 2005. The contribution of commercial fertilizer nutrients to food production. Agron. J. 97(1):1–6. doi:10.2134/agronj2005.0001

Syswerda, S.P., and G.P. Robertson. 2014. Ecosystem services along a management gradient in Michigan (USA) cropping systems. Agric. Ecosys. Environ. 189:28–35.

Vitousek, P.M., R. Naylor, T. Crews, M.B. David, L.E. Drinkwater, E. Holland, P.J. Johnes, J. Katzenberger, L.A. Martinelli, P.A. Matson, G. Nzgiuheba, D. Oijina, C.A. Palm, G.P. Robertson, P.A. Sanchez, A.R. Townsend, and F.S. Zhang. 2009. Nutrient imbalances in agricultural development. Science 324(5934):1519–1520. doi:10.1126/science.1170261

Wortmann, C.S., A.R. Dobermann, R.B. Ferguson, G.W. Hergert, C.A. Shapiro, D.D. Tarkalson, and D.T. Walters. 2009. High yielding corn response to applied phosphorus, potassium, and sulfur in Nebraska. Agron. J. 101:546–555.

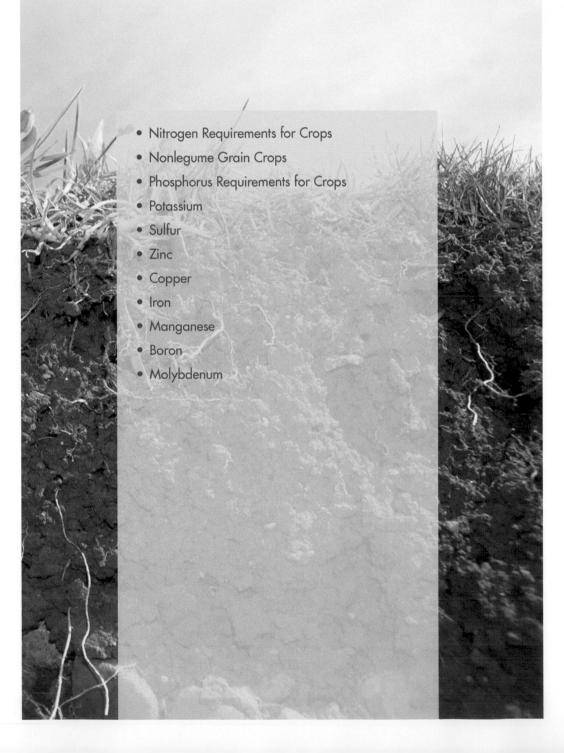

Crop-Specific Nutrient Management

David Franzen

Nutrient management is important for all crop production. Soil samples are obtained and nutrient levels are analyzed. However, the meaning of the soil test results and the nutrient requirements for crops are often very different depending on the crops to be grown. The crops that are grown in a region originate from many places. Some of the nutrient requirements of crops relate to the region in which they were originally grown. Some crops require special nutrient needs as a result of their breeding for special grain or oil characteristics. Some simply require different nutrients at different amounts because of yield potential, quality characteristics, or because of inherent nutrient requirements for the family of plants from which the crop originates.

Nitrogen Requirements for Crops

Nitrogen is required by plants for many reasons. One class of crops, the legumes, can utilize soil N but can also support a symbiotic relationship with N-fixing bacteria.

Soybean

The N-fixing bacteria required for soybean [*Glycine max* (L.) Merr.] is *Bradyrhizobium japonicum*. Soybean is a relatively recent introduction into the United States; however, its cultivation has ancient roots in Southeast Asia, with seeds found in China, Japan, and Korea between 9000 and 5000 yr BP. Seed sizes found at different archeological sites vary between locations, with increasing seed size found gradually over several thousand years. Domestication of soybean tends to be concentrated ~3000 yr BP in the Huanghe–Yellow River basin in China, where a long history of intensive agriculture and population density is evident. Analysis of the evolutionary branches of the N-fixing bacteria responsible for most infection of soybean indicates that strains of *Bradyrhizobium japonicum* in the United States have similar roots of origin as soybean, suggesting that they evolved together (van Berkum and Fuhrmann, 2000).

Abbreviations: AMPA, aminomethylphosphonic acid; DTPA, diethylenetriamine-pentaacetic acid; IDC, Fe-deficiency chlorosis; NDVI, normalized difference vegetation index.

David Franzen, North Dakota State Univ., Dep. 7180, P.O. Box 6050, Fargo, ND 58108 (david.franzen@ndsu.edu).

doi:10.2134/soilfertility.2014.0008

Soybean, when it is grown on a soil for the first time, requires the proper inoculation. Studies examining performance of soybean with and without an inoculant indicate a high reliance on inoculation for greatest yield (Kandel, 2012). Studies on the use of inoculants, even improved strains, show little yield increase as a result of inoculation if the field was previously inoculated when soybeans were last grown. In the 1970s, the concept of source and sink for soybean carbohydrate and nutrition distribution physiologically indicated that early in the growing season, the sink for carbohydrates and nutrients within the soybean was partially directed to the nodules to fuel the production of ammonium N by the bacteria (Keyser and Li, 1992). Later in the growing season, the sink gradually changes from nodule to soybean seed nourishment. The source and sink concept suggested that as the nodules decreased in productivity, supplemental N might increase yield. The most sensational product of the late 1970s was the public announcement of a 20% yield increase with multiple applications of a liquid fertilizer formula (Tietz, 1976). However, most of the experiments by Garcia and Hanway (1976) did not increase yield significantly. Subsequent studies on foliar application of N after seed initiation did not support the announcement. Additional studies using a wide variety of N sources found only a small fraction of experiments had higher yields with post-N application (Mallarino et al., 2001). It is most probable that soybean accumulates N from a variety of sources, including symbiotic N production, soil inorganic N, and accelerated inorganic N release from microorganism activity (Vanotti and Bundy, 1995), so that when the sink changes from nodule to seed, N is mobilized from regions of plant N storage to seed.

Field Pea

The first evidence of pea [*Pisum sativum* L. subsp. *sativum* var. *arvense* (L.) Poir.] (Fig. 1) domestication is ~9000 yr BP in southern Turkey, northern Iraq, Israel, and Palestine. Peas are also found in archaeological remains from Neolithic Greece (7500 yr BP), Bulgaria (6300 yr BP), and the lower Rhine Valley (6200 yr BP). Both Middle Eastern and European evidence of pea is often found with wheat (*Triticum aestivum* L.) and barley (*Hordeum vulgare* L.) grain (Zohary and Hopf, 1973). Ancient pea from the Middle East belongs to the *P. humile* Boiss. & Noe species, which continues to exist as a wild-type at

Fig. 1. Field pea has been domesticated for more than 9000 yr.

present. The modern *P. sativum* cultivars are related closely enough to make genetic crosses.

The relationship between field pea and its N-fixing bacteria *Rhizobium leguminosarum* is relatively strong (Denison and Kiers, 2011). Numerous studies have tested N rate on field pea compared with inoculation alone. In some studies, inconsistent yield increases with N addition were found (Deibert and Utter, 2004), while some studies found little benefit to N addition (Henson, 2004). Experiments in other parts of the world sometimes found that small amounts of N early in the season were beneficial for producing healthy plants capable of supporting subsequent N-fixation relationships. Except for degraded lands with little N mineralization capacity (Erman et al., 2009), most experiments showed only small yield increases with added N. A series of field experiments in Alberta found between 22 and 33% of sites produced yield increases from 6 to 11% with the application of preplant N. Yield increase frequency was greatest with spring soil nitrate levels <20 kg ha^{-1} in the surface 30-cm depth. With the current range of N costs and field pea price structure, the economic return from adding N to pea in most of the major growing areas of the United States would be unlikely.

Lentil

Earliest evidence of lentil (*Lens culinaris* Medik.) (Fig. 2) gathering is ~10,000 yr BP in what is now northern Syria (Zohary and Hopf, 1973). Lentil evidence is common throughout the Middle East in archaeological dig strata dating to ~9000 yr

Fig. 2. Domestication of lentil began ~8000 yr ago.

BP. Larger seeds in great abundance are found in the remains of settlements from ~8000 yr BP, indicating domestication. Lentils then moved to Neolithic Greece and southern Bulgaria. Lentils were cultivated in Macedonia ~7000 yr BP. Lentil become common in Germany by the Iron Age. The original lentil, which is a wild-type today, is *L. orientalis* (Boiss.) Hand.–Mazz., and it grows commonly in shallow rocky soils and soils of Steppe environment. The N-fixing bacterium specific for lentil is *Rhizobium leguminosarum*.

Inoculation of lentil in Canada has increased grain yield in field experiments, but N rates added to inoculated lentil did not increase yield (Bremer et al., 1989). Nitrogen recommendations for lentil in North Dakota do not include preplant N, as previous trials in the state have not shown a response. Only inoculation is recommended in North Dakota for lentil production (Franzen, 2010).

Chickpea

The ancestor to modern chickpea (*Cicer arietinum* L.) is *C. pinnatifidum* Jaub. & Spach, which is a wild-type today and only found in Palestine. Chickpea evidence is first found ~7000 yr BP in present-day Turkey, with some sources citing evidence dating to 10,000 yr BP. By the early Bronze Age, a larger-seed chickpea is found throughout Palestine, indicating domestication (Zohary and Hopf, 1973). The N-fixing bacteria *Mesorhizobium cicero* is native to the Middle East and special inoculation is not required, but in the United States, chickpea seed must

be inoculated for fixation to take place. In the Middle East, isolates of *Burkholderia* spp. also successfully inoculate chickpea and lead to N-fixation (Abi-Ghanem et al., 2012).

Nitrogen fertilization in Pakistan at a rate of 30 kg ha^{-1} increased chickpea yield over use of inoculation alone. In this area near Peshwar, a low rate of N is often required to keep any annual legume healthy enough to produce nodules (Khattak et al., 2006). Contrary to foliar N application studies in soybean, chickpea was responsive to first flower, and 50% bloom application of 30 kg ha^{-1} N as urea solution before the onset of terminal drought increased yield in an Australian greenhouse study (Palta et al., 2005). In a Saskatchewan field experiment, starter N of 30 kg ha^{-1} N as a side-band at planting increased yield in desi chickpea, but had no effect in kabuli chickpea (Walley et al., 2005). Chickpea yield was maximized in another Saskatchewan field experiment using inoculation alone (Gan et al., 2010). Chickpea N recommendations in North Dakota only support inoculation not supplemental N.

Dry Edible Bean

Common bean (*Phaseolus vulgaris* L.) originated in Mexico (Bitocchi et al., 2012) (Fig. 3). Presence of domesticated bean found in Mexico and Peru date to ~10,000 yr BP. However, cultivation of common bean appears only ~2500 yr BP in the Tehuacan Valley of Mexico, 2100 yr BP in the Valley of Oaxaca, and 1300 yr BP in Tamaulipas

Fig. 3. Common bean is generally considered a poor N-fixing legume.

(Kaplan and Lynch, 1999). Common bean is generally considered a poor N-fixing legume with *Rhizobium leguminosarum* biovar *phaseoli* (Bliss, 1993). Although it is also considered a highly permissible legume, allowing fixation for any number of wild-type N-fixing bacteria, including five species of *Rhizobium* (Asadi Rahmani et al., 2011), symbiotic N fixation is inconsistent. Despite its ability to support symbiotic N fixation, common bean is often N deficient without supplemental N.

Studies in North Dakota and Minnesota regarding inoculation and N application to common bean by several researchers showed that three different recommendations strategies may be adopted by growers in the region, depending on the past history of dry bean production (Franzen, 2006). Some growers with well-drained soils with good soil moisture holding capabilities and a long history of dry bean production do not require either inoculation or N application. Other growers produce favorable dry bean yield using inoculation alone. Their soils also tend to be well drained, with good soil moisture holding capabilities and little yield reduction from salinity. The last category of N strategies requires supplemental N, without inoculation, because the soils have problems with salt, poor drainage, are susceptible to seasonal drought, or have poor aeration. In these soils, inoculation is not effective, since a dry bean under stress tends to slough off nodules. The N rates required to produce expected yields are relatively modest ($\leq$80 kg ha^{-1} N). The choice of one of these three recommendations reinforces management of N for not only a crop but also the soil.

Nonlegume Grain Crops

Maize or Field Corn

Maize, or field corn (*Zea mays* L.) (Fig. 4), was domesticated ~10,000 yr BP at a single time from a subspecies of the wild Meso-American grass teosinte, classified as *Z. mays* L. subsp. *parviglumis* H.H. Iltis & Doebley (Iltis and Doebley, 1980). Evidence from starch grains and milling tools found at sites dating to ~9000 yr BP indicate that general corn cultivation was dispersed and expanding after the initial domestication event (Ranere et al., 2009). The area in which the domestication occurred was in a seasonally tropical area of southwest Mexico.

Next to adequate seasonal moisture, N availability is the most important factor for

Fig. 4. Nitrogen availability is one of the most important factors for high-yield commercial corn production.

high-yield commercial corn production. Historically, in the US Corn Belt, N recommendations have been made based on a prediction of yield that was used in a formula similar to the following equation: N rate = (yield prediction in bushels per acre) × 1.2, less N credit from a previous annual legume, N from manure, residual soil test nitrates, or other modifications (Gerwing and Gelderman, 2005).

In Nebraska, soil organic matter, along with expected yield, is included in the N recommendation formula, with soil test nitrate, N from manure, previous crop N credit, and nitrate in irrigation water as major modifications. In North Dakota, western Minnesota, Montana, and part of South Dakota, residual nitrate analysis to 0.61 m (2 feet) in depth is also considered. In North Dakota, failure to include residual soil nitrate greatly reduces the effectiveness of the N rate to corn yield relationship.

The N recommendation formula based on yield prediction has fallen out of favor with most Corn Belt states because it is unrelated to the quadratic response curve that most N-rate studies depict, there is no regard for economics of N rate and corn price, and it assumes that corn grown on all soils have similar response to N.

The maximum economic return to N rate is an economic production function approach that has been adopted by Illinois, Iowa, Indiana, Wisconsin, Michigan, Minnesota, North Dakota, and Ohio and its basic approach has been described by Sawyer and Nafziger (2005). A detailed description of the recommendation system for participating states can be found in Sawyer et al. (2006). In recognition of different soil response to N, some states, including Illinois, Wisconsin, and Indiana, have been categorized by either region, soil texture, or both. The recommendation is based on the original response of corn to N rate for the state or part of the state. The cost of the N and the corn price is considered to provide the greatest profit for the application rate but not the greatest yield. In North Dakota, high clay soils (>50% clay) generally required more N than soils with less clay probably as a result of denitrification in wetter seasons (Franzen et al., 2014). In the regional pooled data, high clay soils also responded differently (Tremblay et al., 2012). In North Dakota, soils with less clay are also subject to leaching losses. Soils with denitrification and leaching potential require a higher rate of N as a preplant rate to achieve similar corn yields as soils that have less risk of N loss. However, a more proper strategy, both economically and environmentally, is split application of N. Nitrogen recommendations (Scharf and Lory, 2006) for Missouri corn growers strongly recommend split application of N to avoid early-season losses of N. Nitrogen recommendations for North Dakota corn growers also strongly recommends split-N application for high clay soils and medium textured soils with leaching potential (Franzen, 2013).

Corn N uptake through the sixth leaf (V6) is only ~10% of total seasonal N uptake. For the next 3 wk, from V6 to tassel, ~50% of the total seasonal N is taken up. When the total N requirement is applied preplant, the N is subject to loss in wetter seasons before the majority of corn uptake. Applying some N preplant to keep the corn healthy until side-dress is important; however, application of substantial N later in the growing season avoids the most common periods of N loss potential.

Corn side-dress N recommendations are beginning to be based on algorithms from active-optical sensors, such as the GreenSeeker (Trimble) or the Holland Scientific Crop Circle sensor (Holland Scientific (Dellinger et al., 2008; Solari et al., 2008, Sharma et al., 2013; Franzen

et al., 2014). An active-optical sensor emits a pulsed light in a code, similar to the technology that opens a particular garage door and not the entire neighborhoods' garage doors. The sensors can be used any time of day because they only sense the light reflected from that emitted. Relationships have been built in several states and countries between the sensor readings and crop yield. Using an N-rich strip, or in some cases an internal standard (Holland and Schepers, 2013), the side-dress applicator would have the algorithm uploaded into the fertilizer controller. The yield predicted from the N-rich strip or internal standard is the yield possible with any rate of N applied at side-dress. Lower yield predicted with lower sensor readings within the field prompt a calculation of yield differences multiplied times the N concentration in corn (1.25%) divided by an efficiency factor to provide the N recommendation (Franzen et al., 2014).

Wheat

Wheat (Fig. 5) was domesticated ~10,000 yr BP in the area known as the fertile crescent of the Middle East from its ancestor Emmer wheat [*T. turgidum* L. subsp. *dicoccon* (Schrank) Thell.]. The primary step in domestication was the selection of nonfragile rachis mutants from Emmer wheat, which allowed more complete harvest and threshing (Peleg et al., 2011).

Nitrogen fertilization strategies for wheat vary depending on the type of wheat to be grown. Winter wheat is the dominant wheat in the United States, with spring wheat and durum wheat (*T. turgidum* L.) dominant in Montana, North Dakota, and northwest Minnesota.

In winter, wheat production—wheat sown in the fall—goes through a period of vernalization. The wheat then begins to grow in the spring, which, in the southern United States, may begin as early as February; the farther north the production, the later dormancy is broken. In the southeast United States, when dormancy breaks, consideration is given to the stand and tiller density before determining N application rates. If the density is below a certain level, an early application of N is applied to stimulate tillering. If the stand is already sufficiently dense, N application is delayed until about the five-leaf stage to avoid N exposure to early rainfall losses. In winter wheat production, protein content of the grain is less important to millers than that of spring wheat to the north, so application of N for yield is the greatest concern for winter wheat

Fig. 5. Nitrogen fertilization strategies for wheat vary depending on the type of wheat to be grown.

growers. For example, in North Carolina (Weisz and Heiniger, 2004), only a small rate of N is applied at seeding: from 11.2 to 33.6 kg ha⁻¹ (20–30 pounds per acre). If in late January to early February there are >538 tillers m² (>50 tillers per square foot), 56 to 78.4 kg ha⁻¹ (50–70 pounds per acre) N should be applied as soon as practical. At growth stage 30 (early jointing), 89.6 to 134.4 kg N ha⁻¹ (80–120 pounds N per acre), less wheat, was applied to stimulate tillering. A tissue test using an in-state technique is recommended in North Carolina to better determine Zadok growth stage 30 N requirements.

Variability in winter wheat protein content tends to make southeast United States wheat less desirable to millers. Investigations into strategies to reduce protein variability in North Carolina concluded that the range of N rates used in the region should be reduced, that only the rates required for optimum economic yield be applied, and to apply spring N at Zadok growth stage 25 (Farrer et al., 2006).

In Kansas, fertilizer N recommendations are for spring application based on predicted yield less a factor based on soil organic matter and other N adjustments including soil test nitrate analysis and previous crop consideration (Leikam et al., 2003).

Development of active-optical sensors for in-season N application was largely based on its application to winter wheat after dormancy break (Raun et al., 2005). Numerous studies have examined the use of active sensors to direct N applications on winter wheat (Li et al., 2009; Butchee et al., 2011; Raun et al., 2002). The use of active-optical sensors to direct wheat top-dress N application is gaining momentum as Oklahoma first led its development and Kansas also has algorithms in place on its

recommendation website N application (Kansas State University, 2015).

Spring wheat and durum wheat production income comes from a total of grain yield and protein content (Wilson and Miljkovic, 2011). In some years, protein premiums are low, or nonexistent; however, in some years, protein premiums may exceed $40 t⁻¹. Discounts for lower protein than the industry's 14% protein standard are in place every year, with the magnitude related often to the magnitude of protein premiums. Therefore, in spring wheat and durum wheat production, adequate N is not only required for higher yield but also to drive protein levels at least >14%.

South Dakota, Minnesota, and Montana support a yield expectation model for N recommendations. In 2010, North Dakota adopted a maximum return to N approach, with consideration of both yield and protein changes with N rate. The North Dakota N recommendation system (http://www.ndsu.edu/pubweb/soils/wheat/) is based on separate N recommendations for soils west of the Missouri River and those to its east. It also segregates out an area in the north–central part of the state characterized by soils with significant soil shale pieces (Redmond and Omodt, 1967) that contain high amounts of mineralizable ammonium (Power et al., 1974). The recommendations also recognized that fields in no-till for greater than six consecutive years have greater macro- and microbiology than conventional till fields and have a higher efficiency of N fertilizer use (Franzen, 2009). Most N recommendations are designed for preplant or at-planting application. The spring wheat and durum wheat is very fast growing, growing from seeding to pollination in 60 d and harvestable in a total of 90 d. Therefore, the opportunity for in-season N application is not great. Wet fields at optimal in-season fertilization stages prevent maximizing yield and protein in some years, while extended dry periods prevent higher N-use efficiency. In areas with more regular rainfall patterns, such as Europe, in-season N application is a normal practice. In addition to preplant N application for spring wheat, anticipated lower grain protein might be increased by the application of ∼33 kg ha⁻¹ (∼30 pounds per acre) of N as urea–ammonium nitrate or solubilized urea immediately postanthesis (Franzen, 2009). The postanthesis application of N for protein enhancement may result in an increase of 0.5 to 1% protein.

Barley

Barley (*Hordeum vulgare* L.) (Fig. 6) was domesticated ~10,000 yr BP in the same area of the Middle East as wheat, pea, and lentil. *Hordeum spontaneum* K. Koch is the wild relative of barley. The origin of barley is in the Fertile Crescent region, now Israel, Jordan, and Iraq (Badr et al., 2000).

Although barley is specifically grown for seed, and N rate is usually based on a yield expectation formula, malting barley N recommendations are based on a more conservative approach. Enough N is required to meet yields supported by rainfall and temperature, but too much N results in higher grain protein and lower kernel size (plump), which are both important to the barley malting industry. Higher grain protein is detrimental to brewing efficiency. Lower plump is also detrimental to the malting and brewing industry. Grain with less than 13% protein is desirable, and higher plump is also desirable. Failure to meet the standards set by the point of sale often results in rejection of a load or substantial discounts down to feed-barley grade (Franzen and Goos, 2007).

In North Dakota, one of the United States' largest volume barley-growing states, a conservative approach to N application is recommended. The state is divided into the western region, with historical higher summer temperatures and lower soil moisture, and the eastern region with higher yield potential, more moderate temperatures, and generally higher soil moisture. Sometimes, such as in 2013, these environments are reversed; however, this weather phenomenon is rare over the past 120 yr. Nitrogen application is recommended for preplant application only, as in-season N application may result in higher grain protein. Nitrogen recommendations are based on expected yield in both Montana and North Dakota. In North Dakota, N rates for expected yield are less in western areas than eastern regions because of the probability that yield might be reduced as a result of lower-than-expected rainfall, higher pollination temperatures, or both.

Sunflower

Sunflower (*Helianthus annuus* L.) (Fig. 7) is a New World crop domesticated in eastern North America ~4000 yr BP (Blackman et al., 2011). Perhaps as a result of the development of sunflower relatively recently in human history and its breeding as an oil seed related

Fig. 6. Barley nitrogen rate differs for malting barley.

mostly to yield, the efficiency of N use by sunflower is relatively high. Nitrogen application increases plant height but may or may not increase yield. Oil-seed sunflower is grown, as the name indicates, for oil content. Increasing N rates increases seed protein and decreases oil concentration (Darby et al., 2013). Currently, N recommendations for most states, including North Dakota, South Dakota, Nebraska,

Fig. 7. Nitrogen use by sunflower is relatively high.

and Kansas, are expected-yield-based formulas. Formulas are modified by soil test nitrate (Nebraska, South Dakota, North Dakota, Kansas), organic matter (Kansas, Nebraska), and in all states by manure and previous crop credits. In contrast to the implied-N-response formula, often, yield expected (kg ha^{-1}) is multiplied by 0.05 less credits.

Recent N-rate experiments have achieved less yield response to N than expected (Darby et al., 2013; Scheiner et al., 2002).

Canola

Canola (*Brassica* spp.) (Fig. 8) was developed in Canada through standard breeding techniques. Its origin is from what Europeans call rape seed, which is also referred to as oil-seed Brassica (*B. campestris* L. and *B. napus* L.) (Gupta and Pratap, 2007). The word rape, comes from the Latin, meaning turnip. The crop probably originated in India, with Sanskrit references to the *B. napus* species ~4000 yr BP. Greek, Roman, and Chinese writings from ~2500 yr BP mention *B. rapa* L. and its medicinal properties. Rape seed was introduced to Europe in the 1300s.

Rape seed was used for industrial purposes mostly as a lubricant. Its use as a food was limited because of its high eruscic acid, which has been shown to cause lipidosis in mice, but not humans, and high glucosinolates, which were highly distasteful. In the 1970s, conventional breeding in Canada was conducted with the goal of achieving a high-quality oil seed for human consumption. With help from a Polish scientist, the goal of zero erusic acid and low glucosinolates was achieved. Canola was the name given to the new subspecies of *B. napus* and *B. campestris*, and canola is now cultivated in Canada, Europe, and the United States. Its oil is considered very healthy for humans and its consumption is surpassed only by palm and soybean oil worldwide.

Nitrogen applications for canola are almost always conducted preplant or at planting. Nitrogen rates in North Dakota are based on a yield-expectation formula with modifications for soil nitrate, manure, and previous crop. The yield expectation formula is limited to 2018 kg ha^{-1} (1800 pounds per acre) canola in the drier, warmer areas of the state to the west, and 2578 kg ha^{-1} (2300 pounds per acre) in the generally cooler and moister areas of eastern North Dakota. Capping N rates at 134.5 kg ha^{-1} (120 pounds per acre) N in the west and 168 kg ha^{-1} (150 pounds per acre) N in the east provided sufficient N to produce healthy, non-N limited plants up to flowering. After the N recommendation for a 2242 kg ha^{-1} (2000 pound per acre) canola crop is provided, higher yields possible with a favorable temperature, adequate moisture, and lack of pests are not N driven but environment driven. With 168 kg ha^{-1} (150 pounds per acre) N, less credits, in North Dakota, close to 4483 kg ha^{-1} (4000 pounds per acre) of canola has been achieved in favorable years (Franzen and Lukach, 2007). The North Dakota experiments support Canadian experiments, where similar results were found (Karamanos et al., 2007).

Sugarbeet

The beet (*Beta vulgaris* L.) (Fig. 9) was introduced to Europe by the central Asian Huns after the fall of the Roman Empire by way of Bohemia from Eastern Europe and Asia. In 1705, Olivier de Serres considered that alcohol might be made

Fig. 8. Canola was developed in Canda through standard breeding techniques.

Fig. 9. Sugarbeet was developed as a source of sugar from the beginning of its cultivation.

from the sugar contained in the beet. Extraction of sugar from a beet was pursued by Margaff, a Prussian chemist, who extracted 5% by weight sugar from beet with saccharine properties similar to cane sugar (*Saccharum officinarum* L.). In 1786, Abbe Commerel published that farmers might feed beet to livestock. Archard, a French scientist living in Prussia revisited the work by Margaff and, encouraged by Frederick the Great, worked out a commercial process to extract sugar from beet. In 1796, the first beet sugar plant was established near Steinau on the Oder River. A rapid increase in sugar production from beet followed the Napoleon era, and by 1878, over 30,000 t of sugar was produced in France from over 500 factories. Beet production was attempted in the 1800s in the United States with little success: first in Pennsylvania in 1839, then in Illinois in 1863, then in Wisconsin in 1866 at Fond du Lac, then in California in 1870 (Ware, 1880). Successful sugarbeet enterprises were eventually established in the Great Plains and in California in the late 1800s, and most centers of production in California, Colorado, Wyoming, Montana, Nebraska, North Dakota, and Minnesota are active today.

The sugarbeet (*B. vulgaris* L. subsp. *vulgaris*) is a variation of the table beet, as is chard [*B. vulgaris* L. subsp. *cicla* (L.) Schubl. & G. Martens]. Wild beet plants are found in the Near East, Mediterranean, Asia Minor, and the Caucasus. Table beet originated as a leafy vegetable in the Mediterranean and spread east in ancient times with a secondary branch developed in the Near East. Descriptions of chard are provided by Aristotle and Theophrastus (Ford Lloyd, 1995). Modern table beet was selected for its swollen root characteristic in Germany in 1558. Sugarbeet for its sugar was developed in the late 1700s (Fischer, 1989).

Nitrogen application for sugarbeet is a balance between too little and too much. Enough N is required to produce a high root yield, but as N rate increases, sugar content decreases (Franzen, 2003). Nitrogen recommendations are designed today in the North Dakota and Minnesota Red River Valley region to support sugarbeet crops from 44.8 t ha^{-1} (20 t per acre) to over 67 t ha^{-1} (30 t per acre). There is only one rate of N recommended for these producers, since it is impossible to accurately predict the final yield preplant. In seasons where high early-season rainfall may result in N loss, sidedress at the six-leaf stage is recommended to provide adequate N, although conservative rates in season are necessary because of the hazard of lower sugar content at harvest from excessive N. Lower N rates are recommended for growers in the Minnesota River Valley and the Southern Minnesota Cooperative region, likely as a result of higher soil organic matter levels and subsequent greater N mineralization than Red River Valley soils. Algorithms for use with active-optical ground based sensors and satellite normalized difference vegetation index (NDVI) imagery are currently being explored, and their incorporation into N management for sugarbeet is expected in the future.

Since the N contained in sugarbeet leaves act as a green manure for subsequent crops in most of the United States, satellite imagery is often used to predict an N credit (Franzen, 2004a; Bu, 2014).

Potato

Potato (*Solanum tuberosum* L.) has been cultivated in the New World for between 7000 and 10,000 yr BP. The greatest diversity in wild potato is in the Lake Titicaca region of Peru and Bolivia, where it was probably first domesticated. Potato was probably domesticated from the wild diploid species *Solanum stenotomum* Juz. & Bukasov, which hybridized with *Solanum sparsipilum* (Bitter) Juz. & Bukasov to form the amphidiploid that evolved and was selected into potato with characteristics that we are familiar with today. Potato grown in ancient beds used techniques that we use today: mounded cultivation, ridges, or raised beds. Storage structures constructed to be naturally cool in ancient times are similar to storage structures today but without the sophisticated electronic controls (Kiple and Ornelas, 2000).

Nitrogen for potato is particularly important because a constant supply is necessary through the growing season or they become susceptible to misshaping and mature at different rates within the field, potentially causing storage issues and possible harvest damage. Different potato cultivars require different rates of seasonal N depending on the N-use efficiency or yield potential of the cultivar. In Colorado, the range of total seasonal N application rates varies from 145.7 kg ha^{-1} (130 pounds N per acre) in 'Russet Nugget' to 224.2 kg ha^{-1} (200 pounds N per acre) for 'Russet Burbank' (Davis et al., 2014). Minnesota recommends that N rate be based on expected tuber yield, with maximum total seasonal N rates for yields >62,768 kg ha^{-1} of 280 kg ha^{-1} of N, less credits, split into at least three application timings (Rosen, 2014). In Idaho, where yields may exceed 67.25 t ha^{-1} (600 hundred weight per acre), total seasonal N rates may be as high as 358.6 kg ha^{-1} (320 pounds N per acre) if preplant soil test levels are near 0 (Stark et al., 2004). The seasonal N rate in all states is always split. In dryland production, such as much of Grand Forks, North Dakota through Crookston, MN, a base rate of about half of total N is applied at seeding, with the other half applied at hilling approximately a month later. Under irrigation, N may be applied several times, commonly with about one-third preplant, followed by one-third at hilling, and the rest applied through the irrigation pivot as directed by petiole sampling for petiole nitrate concentrations. In all of the irrigated potato production areas, including Oregon, Idaho, North Dakota, Minnesota, and other states, state recommendations for critical petiole nitrate levels have been established by University potato researchers to guide growers and their consultant with in-season N timing and rates.

Flax

Evidence of flax (*Linum usitatissimum* L.) (Fig. 10) cultivation dates to ~10,000 BP in the Near East (Allaby et al., 2005). The domestication of flax for fiber and seed is less clear. The cultivated flax is a diploid and the accepted ancestor is pale flax (*L. angustifolium* Huds.), whose origins are from the Tel Abu Hureya region in present day Syria ~11,000 yr BP (Fu, 2011). Cultivated flax and pale flax cannot cross-pollinate. Most recent studies indicate that earliest flax breeding was probably for fiber, with flax oil breeding proceeding a little later in domestication.

Fig. 10. Nitrogen application for flax is usually conducted preplant or at planting.

Nitrogen application for flax is usually conducted preplant or at planting. There is a range of N rates suggested by researchers. In Chile, yield increases to ~1100 kg ha^{-1} were found up to 200 kg N ha^{-1} (Berti et al., 2009). In the Chilean study, lower N rates supported yield at a 20 bushel per acre site, while the higher N rate was necessary for the higher yield sites. The soil in the Chile sites had very low N mineralization rates compared with many soils. In North Dakota, N rates above 89.7 kg ha^{-1} (80 pounds N per acre) often result in lodging and poor harvest recovery; therefore a single rate of 89.7 kg ha^{-1} (80 pounds N per acre) less credits is recommended to growers (Franzen, 2004b,c). In Saskatchewan (Rowland, 2014) the N recommendation is from 39 to 139 kg ha^{-1} (35–85 pounds N per acre) based on soil moisture at seeding.

Phosphorus Requirements for Crops

Soybean

Soybean has a significant phosphorus (P) requirement. Most states in the United States recommend that P be applied at least in soils with medium soil test P values. Research into very high-yielding soybean under irrigation in Nebraska found that increasing P values above levels currently recommended for normal yielding soybean is not necessary to support even higher yields (Specht et al., 2006). Although soybean is not as dependent on mycorrhiza for P nutrition as corn, soybean still relies on mycorrhizal P uptake as a major mechanism of uptake, particularly early in the growing season (Wetterauer and Killorn, 1996; Fixen et al., 1984). Following previously flooded or fallow conditions, up to twice the P normally required for yield may have to be applied to achieve similar yield as in recropped conditions (Fixen et al., 1984).

The method of P application to soybean has been investigated by several researchers. Working in irrigated soybean in Nebraska, broadcast P produced higher soybean yield than a band applied to the side and below the seed (Rehm, 1986). The yield advantage of broadcast to band at a rate of 44 kg ha^{-1} was ~500 kg ha^{-1} of soybean. In a long-term, no-till study in Iowa, broadcast P applications produced similar or greater soybean yield than banded P application (Buah et al., 2000). In Mississippi, fertilizer placement, either broadcast or injected, had little effect on soybean yield in either conventional tillage or no-till (Hairston et al., 1990). In Illinois at Urbana in soils with sufficient initial soil P, application of P increased yields, but placement, either broadcast or deep-banded, had little effect on yield. This study was additionally important because it indicated that deep-banded P is no more efficient than broadcast P (Farmaha et al., 2011). In the central US Corn Belt, it is common in a corn and soybean rotation for the corn to be fertilized for both crops and soybean to feed off of the residual nutrients not taken up by corn. In Iowa, a study on no-till soybean indicated that at optimum and lower soil P levels, fertilizing soybean separately would be advantageous (Buah et al., 2000). One reason why some studies show increases in soybean yield when soil P is in the sufficient range is because of the P response model choice in the past. Until recently, most P response models were based on a linear plateau model. With improved computer and statistical tools, it is possible to investigate quadratic model and exponential models quickly and with little effort compared with using a "two-ruler" approach with gridded paper commonly used in the 1970s. Comparison of soybean P response models in Iowa produced from a large P rate and soil test P and soybean yield database, the exponential model tended to be superior to the other tested models (Dodd and Mallarino, 2005). The linear-plateau model critical soil test P was ~12 mg P kg^{-1}, compared with the exponential soil test P model of ~21 mg P kg^{-1}, using the Bray P1 extractant (Frank et al., 1998). Soybean critical levels for P were less than those of corn in the same study. Similar yield was produced by broadcast P application, banded P at planting, and deep-banded P in another Iowa study (Borges and Mallarino, 2003).

Field Pea

The frequency of yield increase and the magnitude of yield increase for field pea is less than that of other crops, such as soybean. In Alberta, only 19 out of 52 field trials showed a yield increase of ~7%, on average. There was little difference in broadcast compared with banded P in this study, but the authors cautioned that triple super phosphate was used, and that application with monoammmonium phosphate (11–52–0) would move the advantage more to broadcast P (McKenzie et al., 2001). Another Canadian study showed that all soils

did not respond the same. Loam textured soils tended to have a higher yield response to P than clay loam soils. The magnitude of field pea yield response was ~15% (~500 kg ha^{-1} seed yield) with the application of ~135 kg ha^{-1} of 0-46-0 (Karamanos et al., 2003) applied as a side-band at planting. In Montana, pea yields in a P-rate experiment were generally less than the yields in McKenzie et al. (2001) and Karamanos et al. (2003). Fertilizer seed-placed P (0-46-0) increased yields up to the 76 kg ha^{-1} rate (Gan et al., 2004; Wen et al., 2008).

Lentil and Chickpea

Lentil and chickpea have similar utilization and similar requirements for fertilizer P. In an Australian experiment comparing wheat, lentil, and chickpea response to phosphate fertilizer, wheat was the most responsive followed by lentil then chickpea (Bolland et al., 1999). In a Montana experiment with soil test P (Olsen sodium bicarbonate extraction) in the medium to high range, no yield increase resulted from P application to either three lentil cultivars or three chickpea cultivars (Wen et al., 2008). These results are consistent with the strong relationship of both lentil and chickpea to mycorrhiza.

Dry Edible Bean

Recent research into dry bean (common bean) response to P comes from the tropics, where soils have high P fixing potential. Phosphate application recommendations in the United States largely come from trials 30 to 40 yr old that showed that high plant available P was necessary to support higher yields. Recommendations in Colorado stress that P is probably the most limiting nutrient to dry bean production in that state. At low soil test P levels, up to 44 kg ha^{-1} P$_2$O$_5$ is recommended as a side-band but not with the seed. Broadcast is only recommended if the P will be incorporated into the soil (Davis and Brick, 2009). In Nebraska, up to 67 kg ha^{-1} P$_2$O$_5$ is recommended on low P testing soils if broadcast and half of that if side-banded. Application of P with the seed is not recommended (Hergert and Schild, 2013). Fertilization guidelines in many regions recommend P fertilization up to high soil P levels. All recommendations stress side-banding as most efficient, all discourage seed placement, and some indicate that P rates should be higher if P is broadcast and incorporated rather than applied in a side-band.

Maize or Field Corn

Corn has a highly dependent relationship with mycorrhiza. Corn is one of only a few crops that exhibit fallow syndrome (Ellis, 1998). Fallow syndrome is P deficiency in corn caused by planting corn following a bare soil fallow or chemical fallow or following a nonmycorrhiza-supporting crop such as sugarbeet or canola or any plant in the Cruciferae and Chenopodiaceae families. To overcome fallow syndrome, ~66 kg ha^{-1} of P_2O_5 may need to be applied in a band near, but not with, the seed (Sawyer et al., 2011). Broadcast P alone at any reasonably practical rate is usually not enough to overcome the yield depression of fallow syndrome.

Corn often responds to a banded P application, particularly in northern US environments. A North Dakota study (Hendrickson, 2007) recorded a 3000 kg ha^{-1} yield increase with application of 20 kg P_2O_5 ha^{-1} as a row-placed starter. In Illinois, yield increases to banded P are possible (Ritchie et al., 1997), but the frequency of positive responses is greater in northern Illinois than central or southern Illinois (Ritchie et al., 1998). The response to banded P is most often associated with cooler-than-normal springs. The further north the field location, the greater the frequency of cooler planting and poor early-season growth conditions would be expected. In North Dakota, yield increases of over 1500 kg ha^{-1} have been recorded with banded P than broadcast P. Yield increases in Illinois were generally only ~250 to 500 kg ha^{-1} in responsive years. Yield increases in Iowa from starter P application were not seen (Kaiser et al., 2005). A review of literature over several crops suggested that a two-by-two band of concentrated P fertilizer, combined with a broadcast application of P for corn should be strongly considered because the combination enhances P nutrition in corn when growing conditions are good or challenging (Randall and Hoeft, 1988).

Banded P for corn can be applied with the seed or in a side-band. The two-by-two configuration was designed for corn, which puts out seminal roots from the seed at about a 45° angle downward. A two-by-two band is ~5 cm to the side of the seed and 5 cm below, which intercepts the new roots as they elongate into the soil. The separation of >3 cm from the seed allows for greater rates of fertilizer salts and urea to be applied with little danger of injury

to the seed. Some studies indicate that greater urea-N rates than 56 kg ha^{-1} hinder root exploration of the fertilizer band, and the application may not behave as a starter fertilizer, although it will nourish the crop later into the season.

Application of P in a deeper band away from the seed is also a common practice in some areas. The advantage of deeper banding has appeal to those in drier climates, such as the Great Plains, and for growers in a zero-till or modified zero-till, such as strip-till systems. In zero-till, soil P can become stratified or concentrated in the surface 5 cm of soil over time. Although many long-term tillage studies have examined the effects of stratification and found that crops do not suffer from deficiencies because of lower subsoil P in these systems compared with more uniform P distribution in conventional till (Grove et al., 2007), some growers prefer to place P lower in the soil just to make certain of availability in some years. Deeper P placement may also reduce P lost from the land as a result of wind or water erosion, although the tillage systems that result in stratification also reduce soil movement from erosive forces. Recently, residue decomposition in northern, flatter topographic environments, such as in Manitoba, has been found to release soluble P into early spring melt waters and is the largest contributing P source to increased algal bloom in Lake Winnipeg (Flaten, 2013).

There are three principle P fertilization strategies that are used to fertilize corn: (i) maintenance and buildup, (ii) sufficiency, and (iii) a blend of both strategies. The maintenance and buildup approach is used in several central US Corn Belt states (Fernández and Hoeft, 2009; Mallarino, 2008; Vitosh et al., 1995). These regions are characterized by high corn productivity and a low risk for crop failure. The principle is based on a soil having a P test at or above a level that is no longer restrictive to near maximum crop yield. The grower fertilizes to provide enough P for that years' crop and enough extra to build soil P levels over a time period of their choosing to attain the non-P-restrictive test level. For farms with corn yields at or greater than 12,000 kg ha^{-1}, rates of P at or greater than 200 kg ha^{-1} P_2O_5 for a corn and soybean rotation are not uncommon.

The sufficiency approach is more common where corn productivity may be more variable within a field or between fields, and in regions where the risk of crop failure—mostly as a result of periodic drought—is high (Macnack et al., 2014). The principle for the approach is to fertilize for each crop individually, with higher P rates on lower testing soil and lower P on higher testing soil but with no more P added than the crop needs to maximize yield in any given season. These rates may vary depending on soil moisture status at planting time. Usually a row- or near-row-placed starter P application is a part of the overall sufficiency strategy to increase P application use efficiency. A rate of P used in a sufficiency approach for corn yield up to 10,000 kg ha^{-1} is rarely greater than 60 kg P$_2$O$_5$ ha^{-1}. A mixture of both the sufficiency approach and the maintenance and buildup approach is often used in the same region as the sufficiency approach alone (Franzen, 2010; tables and equations). The mixture strategy takes into account the higher risk of crop failure but also recognizes that long-term corn productivity will be reduced is P tests are low. The strategy recommends greater P rates on fields with low P tests, but the rate of soil test increase is usually much less than if the grower elected to use the maintenance and buildup approach. At this time, the decision to use any of the three strategies is made by the grower and is not regulated by any government mandate.

The dependence on high rates of P for high corn yields may be due to the intensive hybridization process and breeding programs that are nearly always located on high P soils and fertilized with P to a high P fertilizer rates. It is noteworthy that despite the dependence on P, the relationship between corn and the mycorrhiza is still strong and corn becomes P deficient when mycorrhiza numbers are limited by fallow or non-mycorrhiza-supporting crops.

Wheat

Wheat has a high requirement for P. Adequate P is necessary for winter wheat to overwinter best in northern latitudes, to stimulate tillering and for high yields. Wheat responds better than many crops to with-seed or near-seed fertilizer banding (Rasz, 1980; Randall and Hoeft, 1988). Unlike corn, wheat recovers well following fallow or following a non-mycorrhiza-supporting crop such as canola or sugarbeet. In parts of the United States and the world, a wheat-fallow system is still used to build soil moisture and accumulate N from mineralization in semiarid, dryland climate systems. Although wheat recovers better than corn after fallow, yield is decreased after fallow if other fallow-enhancing benefits are provided, such as adequate soil moisture and N (Owen et al., 2010). Wheat grown after fallow nearly always yields more

than continuous cropping in semiarid climates, likely as a result of lower soil moisture in continuous cropped fields. The economics, however, often favor continuous cropping in a diverse crop no-till production system (Johnson and Ali, 1982; Kaan et al., 1982; DeVuyst and Halvorson, 2004; Nielsen et al., 2011).

Barley

Barley is similar to wheat in its requirement for P, its preference for seed- or near-seed-placed P, and its seeming immunity from fallow syndrome. Phosphate is used over twice the efficiency in a seed-placed band in barley than a broadcast rate (Rasz, 1980). Adequate soil P is necessary for adequate tillering and higher yields (Fanning and Goos, 1992). In many regions, barley is one of the first crops seeded, since it grows best under cooler conditions and yield is maximized when pollination occurs during cooler, early-summer conditions. Barley is especially responsive to P fertilization and higher P rates when planted into colder soils (Power et al., 1963).

Sunflower

High soil P levels and high rates of P fertilizer have not been found necessary for high sunflower yields. Fallow does not appear to induce P deficiency in sunflower as it does in corn despite the high reliance on mycorrhiza. Sunflower will respond to P fertilizer if soil P levels are low. In an Australian study, two of seven sites had increased yield as a result of P application in soils with Colwell P extract levels below 20 mg kg^{-1}, which would place the value in the very low P range for wheat. A Colwell P critical level for wheat would be 40 mg kg^{-1} (Peverill et al., 1999). Forty site-years of N- and P-rate experiments were conducted in western Nebraska during 1993 and 1994. There was no effect of P rate on sunflower yield at any location. Most of the sites had medium to high soil P levels, but some sites were classified in the low soil P range (Geleta et al., 1997). In Turkey, in a low P testing soil with pH 7.6 and 4.4% free carbonate, sunflower yield was increased ~1.5 t ha^{-1} with 50 kg ha^{-1} of P as triple super phosphate. Oil content also increased with P application. Yield and oil content was also increased about the same amount with inoculation of an acid-forming *Bacillus* strain, which may have released occluded P, although plant P uptake was not included in the publication

(Ekin, 2010). Treatments included both fertilizer P, and the *Bacillus* tended to produce greatest sunflower yield and oil content. Similar yield responses were recorded from both seed-placed P and broadcast P applications. New North Dakota sunflower fertilizer recommendations state that zero P is required, based on results of 40 recent field P rate experiments in the state (Franzen, 2016).

Canola

High soil P and especially near-seed-placed P is necessary for high canola yield. Canola is a nonmycorrhizal crop. Perhaps it is the necessity of the crop to gather P from the soil without the aid of mycorrhizal fungi that make canola so responsive to banded near-seed P (Rasz, 1980). Other broadleaf mycorrhiza-supporting crops do not respond with particular vigor to seed-placed P. The only major crop that is similar to canola in its preference for seed-placed P is sugarbeet, another nonmycorrhizal crop. In contrast, a study at Carrington, ND, did not observe a yield increase from P in-furrow application compared with broadcast or midrow band application. There was a 50% seed yield increase at one site as a result of the application of 40 kg ha^{-1} P$_2$O$_5$ and a 25% seed yield increase to a similar rate at a second site with a 10 mg kg^{-1} Olsen extract beginning P soil test (Hendrickson et al., 2008). Results of 14 yr of N- and P-rate studies on Canola showed that canola yields increased ~10% with broadcast applications to 92 kg P ha^{-1} (Nuttall et al., 1992). Banding near or with the seed is usually superior to broadcast or deep-banded P (Nuttall and Button, 1990).

Sugarbeet

High soil P is necessary for highest sugarbeet tonnage yield and sugar yield. Sugarbeet responds particularly well to seed-placed P (Sims and Smith, 2001) In many fields, a modest seed-placed P rate of ~10 kg ha^{-1} P$_2$O$_5$ can substitute for 50 kg ha^{-1} P$_2$O$_5$ applied broadcast. Sugarbeet growers in North Dakota and Minnesota have largely adopted the seed-placed P approach and only apply broadcast P to fields that have lower soil P levels; but even then, some seed-placed P is usually applied.

Potato

Phosphorus nutrition is important for high yielding, profitable potato production (Rosen et al., 2014). Final yield is a product of tuber set, tuber growth rate, and duration of tuber growth, which are all related to P nutrition. Tuber quality, expressed as specific gravity of potato tubers, is increased when P is added to low-P soils. Phosphorus deficiency can also result in increased susceptibility to common scab (*Streptomyces scabies*), verticillium wilt (*Verticillium dahlia/V. albo-atrum*), and late blight (*Phytophtera infestans*). Potato production requires large amounts of available phosphate. The critical soil test level for P in potato is higher than most row crops, such as corn or wheat. For example, Bray P critical levels for P for corn or wheat are usually around 50 mg kg^{-1}, but for potato, the levels range from 51 (Minnesota) to 200 mg kg^{-1} (nonsands, Wisconsin). Unlike responses in many crops, P fertilization is recommended in most regions even when P tests are at the critical level, since small yield and quality gains are almost always profitable from P fertilization at rates 50 to 100% of expected crop uptake. An Idaho study recorded total potato yield increasing up to, and according to the regression model constructed, >300 kg ha^{-1} P$_2$O$_5$ broadcast rate. Most studies comparing broadcast with various configurations of concentrated bands of P concluded that banding was superior, in particular a band that is 5 cm away from the seed piece in both directions and even with the seed piece placement depth. Studies recording equal or superior yields with broadcast were influenced by placing the banded application too close to the seed and reducing stand and early plant health. In-season applications of P, usually through fertigation, have usually not been effective in increasing potato yield, and in some studies, yield was reduced. One factor that influences effectiveness of fertigated P in-season is rooting depth. Rooting very close to the soil surface increases P uptake from fertigation, while deeper rooting depth relative to the soil surface decreases or prevents P uptake from a surface application.

Flax

Flax is highly reliant on mycorrhiza for its P nutrition. Numerous P-rate studies in North Dakota and Manitoba found no response to P from any rate of P. In a study comparing flax seeded following canola (a mycorrhiza nonsupporting crop) and wheat, residual P increases from previous canola P rates increased flax yield, but additional P applied directly before flax did not result in a flax yield increase. After wheat, no yield differences as a result of soil P differences from P fertilization of the previous crop or P applied for flax were recorded. Flax is the only crop grown in North Dakota for which a no-P recommendation is given (Franzen, 2004b,c). When P is applied, mycorrhiza activity is suppressed. When no P is applied, mycorrhiza is able to mobilize whatever P is required by the flax. Flax should not be seeded after fallow or after a crop or cover crop exclusively in the Brassicaceae and Chenopodiaceae plant families.

Potassium

Most crop responses to potassium (K) have been calibrated to soil test indices. The extractant used in the analysis varies with region. The K test results can vary with method (Barbagelata, 2006) and with time of the season (Franzen, 2011). Although the prediction of whether a crop requires K for economic yield enhancement varies with the critical soil test level for the crop, the soil sampling and testing methods used for the prediction are similar for depth, timing, and the analysis. Some illite- and vermiculite-dominated soils also have high K-fixing power and require a different extraction method and rate structure. The approach to fertilization in the United States may be based on a maintenance and buildup approach, or a sufficiency approach, previously detailed in the Maize and Field Corn Phosphorus section of this chapter. Unlike with P, soils with relatively low actual cation exchange capacity, approximately <10 cmol kg^{-1} soil, do not have the capacity to build soils more than ~100 ppm. In these soils, application based on crop response is recommended since buildup to critical levels for some crops is impossible. In the United States, some recommendations—usually not University based—recommend K rate based on a balance of cations rather than the K test itself. Most soil fertility scientists associated with US Land Grant Institutions and many soil fertility scientists worldwide reject this approach based on empirical evidence to the contrary (Kopittke and Menzies, 2007).

Soybean, Field Pea, Lentil, Chickpea, and Dry Edible Bean

In soybean and the other annual legumes, yield increases to K when soil tests are near or above 100 mg kg^{-1} K are rare. In nearly all studies,

broadcast K is similar or superior to banded K placement (Buah et al., 2000; Mallarino et al., 2001; Slaton et al., 2005). In Idaho, recommendations for K are only provided if soils have <75 mg kg^{-1} K for chickpea (Mahler, 2005). North Dakota recommendations only provide K fertilizer recommendations in dry bean for soils with K tests lower than 80 mg kg^{-1} (Franzen, 2010). In-season foliar K application in Missouri high-testing K soils increased soybean yield ~5% in a moist-soil environment, but decreased soybean yield when the soil conditions were dry in another year. In 48 soybean foliar rate trials in Iowa, there were no yield increases as a result of treatment (Haq and Mallarino, 1998). Foliar fertilization of common bean can also decrease yield, as evidenced in some experiments (Neptune and Muraoka, 1977).

Maize or Field Corn

Corn is highly responsive to K if soil test levels are low. Deficiency symptoms are often seen when soil test levels fall below 150 mg kg^{-1}. Yield responses are often recorded when soil test levels are >150 mg kg^{-1} when corn is raised in no-till and ridge-till systems (Mallarino et al., 2001). In Iowa, corn yield increases from K application on soils with <150 mg kg^{-1} K test ranged from 1100 to 1400 kg ha^{-1}.

Broadcast K produced similar yield compared with deep-banded or seed-placed K in conventional tillage, whereas deep-banded K was superior to broadcast or seed-placed K in no-till and ridge-till systems (Mallarino et al., 2001). In South Dakota, in a no-till field, a study on soils with 122 mg kg^{-1} K, corn yield increases over 3700 kg ha^{-1} were recorded with application of 132 kg ha^{-1} K_2O (Gelderman et al., 2002). In Nebraska, where no-till and ridge-till corn production is rare, a study in high-yield corn environments (>14.5 Mg ha^{-1}) found that even in higher yielding corn soils, responses to K were not seen with soil test K levels above 125 mg kg^{-1} (Wortmann et al., 2009). In no-till and strip-till (zone-till) systems in Ontario, near-seed banded K was more effective than broadcast K when soil test K levels were <125 mg kg^{-1}. The critical level for K was ~150 mg kg^{-1} for the soils in this study (Vyn and Janovicek, 2001). Some of the inconsistency in critical K soil test levels between states and regions is the nature of the soil test procedure for K. Iowa had an early (before 1995) history of using a field moist soil test. Certainly in some years, field-moist was near to air-dry soil, but in most years, some moisture was present in a soil sample. For a period of time between 1995 and 2010, Iowa reverted to an air-dry system, with 150 mg kg^{-1} as a general critical K level with 1 M ammonium acetate extractant. Recently, because of discomfort with the relationship of the dry soil test with K response data, Iowa again recommends the moist K test (Mallarino et al., 2011). The field-moist test provides a greater relationship with K response in corn than the air-dry soil-based test. The critical level for corn K response using this system is ~100 mg kg^{-1}. Some other regions are now evaluating the field-moist test compared with the air-dry K soil test. Recent studies in Minnesota indicate that in ridge-till systems, the critical air-dry K soil test should be greater than that for conventional tillage systems (Rehm, 1995), although in other studies, no difference was found between tillage systems (Rehm and Lamb, 2004; Fernández and Schaefer, 2012)

Wheat and Barley

Wheat is only responsive to K application if soil test K level is low. In North Dakota, the critical level is 100 mg kg^{-1}. In Alberta, Canada, the critical level is 125 mg kg^{-1}. Both of these critical levels are lower than that of corn in many regions. In other crops, such as potato, application of a small amount of K as a starter or broadcast continues to increase crop yield. However, in wheat, application of row-placed K is not helpful and is sometimes harmful if K soil tests are greater than the critical level (Karamanos et al., 2013). In contrast, starter K applications were found to be helpful in barley ~40% of the time, and most of these yield increases were not associated with soil chloride levels but most often were associated with disease suppression (Karamanos and Flore, 2000). In a hydroponic study (Mullison and Mullison, 1942), low K levels resulted in poor barley early plant growth and poor general plant health. The K source in this study did not include chloride.

Sunflower

Very little recent work has been conducted on the response of sunflower to K fertilizer or establishing K levels for sunflower. The work that has been conducted indicates that K levels similar to that needed by corn (150 mg kg^{-1}) would not be excessive, but K levels similar to grow wheat (100 mg kg^{-1}) would probably not be too low. Many research articles originating outside of the United States on sunflowers and K application do not include beginning soil test K levels and other important soil nutrient factors for consideration.

Canola

Canola is generally unresponsive to K unless the soil test levels are very low (<100 mg kg^{-1}). Even at very low levels, only low rates of K fertilizers are required to alleviate K deficiency (Ontario Ministry of Agriculture, 2014). According to work in Australia, requirements for K in canola are similar and perhaps slightly greater than K rates for wheat (Brennan and Bolland, 2007).

Sugarbeet

Sugarbeet is very responsive to K fertilization when K availability is low (Moraghan, 1985; Hergert, 2010). It is possible for sodium to substitute for K (Shepherd et al., 1959; Wakeel et al., 2010), but sodium can cause serious soil physical problems (US Salinity Laboratory Staff, 1954), so it is not a serious commercially-used substitute for K. In a western Minnesota study with a starting K soil test of 46 mg kg^{-1}, (Cattanach and Overstreet, 2006) tonnage yield and recoverable sugar were increased by 50% by the application of 180 kg ha^{-1} K$_2$O. The K response studies with sugarbeet are not nearly as plentiful as those for N and P, probably because sugarbeet production is usually conducted in drier environments with soils at least originally rich in K, such as the US Great Plains and the valleys of Washington, California, Montana, and Wyoming (Doxtator and Calton, 1950; Fixen, 2002). Although K is an impurity that must be removed from the sucrose stream during processing, current methods are able to accomplish this task without increasing costs even if K fertilization or availability is excessive.

Potato

Large rates of K are applied regularly for potato production. Amounts typically applied in the United Kingdom are over 200 kg K$_2$O ha^{-1} (Dampney et al., 2011). In the United States, the rate for 5 t ha^{-1} potato is typically 180 kg ha^{-1} in Colorado (Davis et al., 2014) and 260 kg ha^{-1} in Wisconsin (Kelling et al., 2002). Idaho recommendations are based on both soil test exchangeable K and yield expectations. For a soil test of 100 mg kg^{-1} K and a 5 t ha^{-1} yield 330 kg ha^{-1} K$_2$O is recommended. Application method of K is not important, whether it is broadcast, band, or a combination approach (Kelling et al., 2002). Potassium application can increase yield and quality of potato if the rate is not higher than the crop requires. Application much less than required results in lower yield, quality, and disease susceptibility, while application at rates much greater than required for yield increase results in lower quality potato (Westermann et al., 1994).

Flax

Most flax fertility guides recommend a rate of K for flax, but few studies have reported flax yield increases from the application of K fertilizer. The rates of K recommended are based on a combination of flax seed removal of K through harvest and soil test. An exchangeable K level of 150 mg kg^{-1} in North Dakota and the Canadian Prairie Provinces is the critical soil test level (Franzen, 2004b,c; Rowland, 2014).

Sulfur

Canola and Other Crops

More than most crops, canola has a special requirement for S. All crops require S, however, application of 22 kg ha^{-1} sulfate-S increased yield of canola in one North Dakota study from ~33 to over 1800 kg ha^{-1}, which is a far greater yield increase than one would expect with any other crop (Halley and Deibert, 1996). A review of S application in canola in the US Northern Plains and in the Canadian Prairie Provinces is provided in Franzen and Grant (2008). Response of other crops besides canola is also discussed in the review.

Maize or Field Corn

Since 2005, S deficiency has become more important to corn in the United States than had previously been reported. Researchers in Iowa had periodically conducted survey S-rate trials before 2005 and had never found a responsive site. From 2005 through present, responses to S in corn became increasingly more common (Sawyer et al., 2012). Sulfur deficiency

has also been reported in North Dakota, usually in sandy loam or coarser, although in very wet spring seasons, S deficiency has also been observed in high clay Fargo silty clay loam soils (fine, smectitic, frigid Typic Epiaquerts) with organic matter >5%. The reason for the S deficiency in coarser textured soils is most likely leaching. However, the probable reason for S deficiency in high clay, high-organic-matter soils is most likely a continuous, slow downward movement of water, which prevents both S mineralization from organic matter and also capillary movement of sulfate-dominated groundwater to the surface (Franzen, 2014). Use of an N-sufficient area within a field has been shown to help indicate a S deficiency using an active-optical sensor. In a S-sufficient field, the N-sufficient area will provide a higher NDVI (red NDVI or red-edge NDVI) than the surrounding area of corn. However, if S deficiency is present, the N-sufficient area will often provide a lower NDVI reading than the rest of the field as a result of enhanced S deficiency in the N-sufficient area. If this phenomenon is experienced, a sulfate or thiosulfate application should be applied to the field as soon as practical. The N side-dress activity based on an active-optical sensor should be delayed at least 1 wk following the S application, preferably following rainfall (Franzen et al., 2014).

Soybean

Although reports of S deficiency in corn have been common, the trials reporting S deficiency in soybean have been far fewer (Sawyer et al., 2012). Multiple S-rate trials using several rates, materials, and methods were investigated in Minnesota (Kaiser and Kim, 2013). Only one site with organic matter concentration <20 g kg^{-1} showed a positive response. The conclusion was that a S response in soybean was possible, but only in low-organic-matter soils under the right conditions.

Sugarbeet

An excellent review of S and micronutrient research in sugarbeet is provided in Christenson and Draycott (2006). There were no S responsive studies in sugarbeet noted in the review. A study in England in 2010 on six sites with loamy sand textures, where responses to S in previous crops had been recorded, found S response at five of the six locations of an average 0.56 t ha[-1]. The authors recommended that sugarbeet be amended with S if grown on similar soils.

Zinc

North Dakota is a state that has a large area of Zn-deficient soils (Franzen et al., 2006) based on the diethylenetriamine-pentaacetic acid (DTPA) soil availability analysis (Lindsay and Norvell, 1978; Whitney, 2012) and its interpretation. Even with relatively low soil Zn levels, only corn, flax, potato, and dry edible bean have shown a response to Zn in the state (Franzen, 2010). However, worldwide, there have been responses to Zn in many crops. Although most legumes, with the exception of dry edible bean, are not responsive to Zn in North Dakota or the Prairie Provinces of Canada, pulses are regularly Zn deficient in India (Tiwari and Dwivedi, 1990). In one study that compared the relative response of eight crops for susceptibility with Zn deficiency, lentil was most susceptible, followed by chickpea, pea, wheat, flax, mustard (*Brassica* spp.), barley, and oat (*Avena sativa* L.). The central Anatolian Plains of Turkey is one of the most recent examples of a region reporting widespread Zn deficiency in wheat. Wheat yields were increased six- to eightfold with the application of Zn. The DTPA–Zn critical level found in the region was 0.4 mg kg[-1] (Cakmak, 2008).

Soybean

Response of soybean to Zn fertilization is rare in the United States. Zinc fertilization has increased soybean yield in responsive soils in other parts of the world, however. Substantial yield increases in New South Wales, Australia, were recorded by Rose et al. (1981) through foliar application on low-Zn testing soils before flowering.

Field Pea and Lentil

Although several fertilizer recommendation guides around the world indicate that field pea would benefit from the application of Zn if soil test levels were low, little field evidence can be found that substantiates a recommendation. Field pea is particularly reliant on mycorrhiza for its Zn nutrition (Ryan and Angus, 2003).

Chickpea

Several studies have reported yield increases as a result of Zn fertilization in Zn-deficient soils in India (Ahlawat, 1990). Chickpea yields were increased with Zn fertilization at one site in South Australia and two sites in Pakistan (Khan et al., 2000).

Sugarbeet

Sugarbeet sometimes responds to Zn application. In a Red River Valley study in Minnesota, Zn had no effect on sugarbeet yield. A multiyear study in Wyoming resulted in yield and quality increases in 2 of 3 yr, with the seed-placed Zn sulfate treatments being the most effective. Broadcast Zn sulfate also increased sugarbeet yield in 1 yr, although not as much compared with seed-placed Zn sulfate (Stevens and Mesbah, 2005). The 2 yr in which Zn application increased sugarbeet yield, site DTPA Zn was 1 mg kg[-1] or below, whereas the year in which no response was recorded, the site DTPA Zn was 1.7 mg kg[-1], which is considered high. Nebraska sugarbeet recommendations indicate that no response to Zn should be expected unless the DTPA Zn test was 0.5 mg kg[-1] or below, and even with low Zn tests, the probability of Zn response is not high (Binford et al., 2000). Christenson and Draycott (2006) cite one study from Iran with a 10% sugarbeet yield increase as a result of Zn application, but no positive studies were found from North Dakota, Minnesota, or Michigan sugarbeet growing areas. Sims (2006) treated a loamy fine sand soil in Minnesota with Zn and found no yield or quality increase in sugarbeet.

Sunflower

In Iraq, foliar sprays of Zn sulfate applied at bud stage to sunflower grown on a soil with a DTPA Zn level of 0.38 mg kg[-1] increased seed yield (Shaker and Al-Doori, 2012). In Pakistan, soil-applied Zn alone did not increase sunflower yield in a soil with a DTPA Zn level of 0.26 mg kg[-1] (Khan et al., 2009). Responses of sunflower to Zn fertilization have not been found in the USA.

Potato

Zinc deficiency in North Dakota was first identified on potato in 1949 (Hoyman, 1949). Land leveled in preparation for irrigation was particularly susceptible to Zn deficiency (Grunes et al., 1961). Most states in the United States recommend Zn application to potato when DTPA Zn levels are <1 mg kg^{-1}.

Flax

Chlorotic dieback in flax was identified by Flor (1943), but he did not relate the condition as a Zn deficiency, rather a condition of soils with high lime content. Chlorotic dieback in flax was further described as Zn deficiency by Zubriski (1964) and later reinforced by Moraghan (1979, 1980, 1982). Soils with DTPA Zn levels of ≤1 mg kg^{-1} are regularly supplemented with Zn.

Copper

The most common Cu deficiency problem in many crops has been in organic peat and muck soils. The application of Cu as a benefit to crops has long been reported (Harmer, 1946; Lucas, 1945).

Copper deficiency in mineral soils of legumes is rare. General Australian recommendations suggest that Cu deficiency might be a problem in field pea, but there is little evidence in the literature that supports that possibility.

Wheat and Barley

Copper deficiency is common in parts of Australia (Reuter, 2007). Diagnosis of widespread Cu deficiency and its correction through foliar Cu sprays on wheat was reported in 1975 by King and Alston (1975). Copper deficiency effects on disease susceptibility and wheat pollen sterility were further described by Graham and Nambiar (1981). Copper deficiency has also been reported and successfully remediated with Cu fertilization in spring wheat and barley in Alberta and Saskatchewan (Kruger et al., 1985; Penney et al., 1988).

A low frequency of spring wheat and durum (*Triticum turgidum* L.) responses to Cu fertilization was also found in North Dakota soils (Franzen et al., 2008); however, the responses were confined to low-organic-matter, deep, sandy soils. In North Dakota, increases in yield at some sites were recorded, and decreases in Fusarium head blight occurrence and severity were also found, but disease was not decreased enough to substitute for a fungicide application.

Iron

Iron deficiency is nearly always associated with the presence of soil carbonates and high pH. High pH reduces Fe solubility, whereas soil carbonates tend to neutralize acidity near roots, rendering Fe-reducing substances secreted by roots ineffective (Franzen, 2013). Wet soil conditions increase the concentration of bicarbonate in soils, which increases Fe-deficiency chlorosis (IDC) (Inskeep and Bloom, 1986).

Soybean

Soybean is particularly susceptible to IDC. In North Dakota, a high percentage of acreage is affected by IDC in wetter years (Franzen, 2013). Soil carbonates become soluble, rendering Fe-reducing substances ineffective at transforming Fe^{+3} to Fe^{+2}, making Fe about a trillion times less soluble. Wet soil conditions, cool soil temperature, presence of soluble salts (Franzen and Richardson, 2000), ill-advised Mn application, and less tolerant varieties increase the occurrence and severity of IDC in soybean. Application of ortho-ortho EDDHA with or near the seed can alleviate IDC. Seeding in wider rows with less distance between plants, seeding using a cover crop to reduce soil moisture and take up excess nitrate, and especially using a more tolerant variety are cultural strategies that decrease IDC. If the spatial hazard of IDC were known, seeding a high-yielding intolerant variety in the non-IDC-susceptible areas of a field and an IDC-tolerant variety in the IDC susceptible area of the field would be expected to maximize yield in the field (Helms et al., 2010).

Lentil, Field Pea, Chickpea

Lentil tends to be susceptible to IDC in calcareous soils, and there is a wide difference in susceptibility to IDC with variety. In a varietal screening study with 3512 cultivars originating from 18 countries, 16.9% of the selections showed IDC in a Calcic Rhodoxeralf (calcareous) soil in Syria. Cultivars from the original lentil cultivation countries (Syria and Turkey) had the smallest percentage of susceptibility. In Indian cultivars with severe deficiency, yield reductions of up to 47% were recorded. Iron deficiencies are common in Australia production in calcareous soils of the Victorian mallee (French and White, 2005). Yield losses in chickpea in Syria and Lebanon to IDC have been estimated at ?44% and from 24 to 50% in India (Saxena and Sheldrake, 1980; Sakal et al., 1987; Yusuf Ali et al., 2002). Considerable possibility for genetic improvement for chickpea tolerance to IDC appears likely (Toker et al., 2010).

Dry Edible Bean

Dry bean is not nearly as susceptible to IDC as soybean; however, IDC is possible when soil carbonate levels are high, soils are wet, and soluble salt concentration or another stress condition has an adverse effect on growth. Iron deficiency in dry bean has been recorded in Nebraska in soils with pH $\geq$ 7.5 (Hergert and Schild, 2013) and in Wyoming in calcareous soils (Stevens and Belden, 2005). Considerable differences to susceptibility between dry bean cultivars exist (Ellsworth et al., 1997).

Maize or Field Corn

In western Nebraska, some soils have a pH > 8.5, and IDC in corn is common on ~0.4 million ha. An Fe-fertilizer-rate study was conducted using seed-row and foliar application of different products at different rates. Most Fe treatments improved yield in soils with pH > 8, but not at sites with pH < 8. A combination of tolerant variety and Fe-sulfate amendments increased corn yields most (Hergert et al., 1996). A banded application of ferrous sulfate provided alleviation of some of the stress from IDC in similar soils in Kansas (Godsey et al., 2003). Corn in North Dakota, where soil pH is seldom >8.5, has not been recorded except under very high soluble salt conditions. Under high soluble salt conditions, the yield is most reduced because of the salt with IDC as a secondary concern.

Wheat, Barley, and Oat

Iron-deficiency chlorosis in small grains is usually the result of choosing a nontolerant cultivar with respect to Fe deficiency. Berg et al. (1993) report a large area of IDC in winter wheat, with the cultivars dominated by those with low tolerance to IDC. Application of foliar Fe sprays to the wheat resulted in alleviation of IDC and doubling forage yield in remedial studies. There are considerable cultivar differences

in tolerance to Fe-deficient conditions between wheat cultivars (Rengel and Romheld, 2000). Iron-deficiency chlorosis is also common in barley in calcareous soils. The severity of IDC in barley is increased with high soluble salts (Yousfi et al., 2007).

Canola

There have been no reports of Fe deficiency in canola reported in the literature.

Sunflower

Iron-deficiency chlorosis has been reported in Spain where sunflower genotypes exhibit different responses to treatment with Fe-EDDHA (Alcantara and de la Guardia, 1987). There have been no reports of IDC or response of sunflower to Fe fertilizers in North Dakota.

Sugarbeet

Sugarbeet is generally thought to be a Fe-efficient plant, and therefore, references to Fe deficiency in sugarbeet are rare. In sandier soils at the edge of the Red River Valley in North Dakota and Minnesota, yield increases with both broadcast ferrous sulfate granules and seed-placed ortho-ortho EDDHA have been reported (Overstreet et al., 2007). The yield differences were recorded at multiple sites and the plots did not exhibit chlorotic conditions.

Potato

Even though substantial potato production occurs in North Dakota and Minnesota soils with pH up to 8.5, no Fe-deficiency symptoms or response of potato to Fe fertilizers has been reported. Articles regarding Fe deficiency in potato induce the deficiency through manipulation of hydroponic element mixture or use of very sandy greenhouse pot mixtures.

Flax

Although not common in North Dakota, IDC has been reported in small areas within fields that with high free carbonates under wet soil conditions (Franzen, 2006). Iron-deficiency chlorosis has also been reported in wider areas on calcareous soils in the Middle East (Esmail et al., 2014). Highly calcareous soils in Australia have also experienced IDC in flax and have been successfully treated with Fe foliar sprays

(Bakry et al., 2012). Flax was found to accumulate toxic levels of Mn when soil is low in available Fe (Moraghan, 1979). Applications of Fe-EDDHA eliminated the Mn toxicity effect.

Manganese

Soybean

Manganese has long been identified as deficient in muck, peat, black sands, and depressional soils with pH > 6.2 in Michigan (Vitosh et al., 1995; Boring and Thelen, 2010), northwest Indiana (Brouder et al., 2003), coastal plain soils in Virginia (Alley et al., 1978), certain soils in Ohio (Kroetz et al., 1977), and a small number of soils in Illinois (Mulvaney and Pendleton, 1967). Supplemental Mn has been recommended for a number of crops including soybean, corn, sugarbeet, wheat, and oat.

Glyphosate Controversy with Induced Mn Deficiency in Soybean

Huber et al. (2004) offered that in their study area in northern Indiana, there was evidence of glyphosate, inducing Mn deficiency in soybean. Glyphosate-resistant cultivars were more deficient in Mn compared with nonglyphosate resistant cultivars. Addition of Mn fertilizers to the glyphosate application resulted in 10 to 50% weed control reduction than no Mn fertilization. Huber et al. relates that soybean treated with glyphosate induces chlorosis, which, in Indiana and the surrounding region, would be interpreted as Mn deficiency. Several studies to determine the nature of the glyphosate effect were conducted by Rosolem et al. (2010). Nutrient solution Mn kinetic studies and greenhouse studies were conducted to compare soybean uptake and metabolism of Mn. When glyphosate was added to nutrient solution at low rates, soybean chlorosis was observed a few days after treatment, but analysis of plants showed that uptake of Mn was not disrupted. The chlorosis was unrelated to Mn nutrition. One possible explanation of yellowing in soybean following glyphosate application is the formation of aminomethylphosphonic acid (AMPA), which is a physiological transformation of glyphosate in glyphosate-resistant soybean. aminomethylphosphonic acid is known to cause yellowing in soybean. Its effects when soybean is treated directed with AMPA may last through 14 d after treatment, but the effects pass by 28 d after treatment. Current Mn recommendations in glyphosate-resistant soybean are not different in Indiana and Ohio than for non-glyphosate-resistant soybean. However, authors of recommendations strongly encourage growers to monitor crop tissue analysis, since Mn deficiency and yellow flash from glyphosate application appear very similar.

Lentil, Field Pea, Chick Pea, Dry Edible Bean

Of these four crops, Mn deficiency is most common in field pea. It was first identified in southern England and carries the name marsh spot (Glasscock, 1941). If field pea develops the deficiency early in the season, yield decrease can be expected, but more importantly, the pea develops a necrotic spot in the center of the cotyledons, resulting in much lower market value if destined for human consumption (Biddle and Cattlin, 2007). Dry bean yield can be decreased if Mn deficiency is severe. Severity of Mn deficiency usually occurs as a result of cultivation in muck or peat soils or in soils with low native Mn and pH > 7.

Wheat, Barley, and Oat

Manganese deficiency is the most common micronutrient problem in Michigan. Winter wheat can be affected by Mn deficiency if it is not fertilized with a Mn fertilizer. Manganese deficiency is also common in Australia, and plants affected are pale, stunted, and the leaves are limp (Government of Western Australia, 2015). Recently, Mn deficiency has become more prominent in Punjab, India. Field demonstrations of foliar Mn amendments are being given to local farmers to help them deal with the problem (Sidhu et al., 2012). Amelioration of aggravated Mn deficiency in wheat crop through field demonstrations in light-textured soils of Punjab. Wheat has been so deficient of Mn in Tanzania that the wheat died (Hoyt and Myovella, 1979).

Taylor and Sylvester (2011) classified barley as the most Mn-sensitive small grain grown in Delaware. They noted that spring N application on Mn-deficient barley often caused the plants to die.

Manganese deficiency in barley is widespread worldwide and is common in Australia, the United States, China, Europe, and Scandinavia (Husted et al., 2009).

Oat is also susceptible to Mn deficiency and, in Ireland, has been classified as the most susceptible small grain (Gallagher and Walsh, 1943). Manganese deficiency in oat was first classified as gray speck disease but was later recognized as Mn deficiency. Manganese problems in oat have been found in Canada (Timonin, 1947), Europe, the United States, and Australia.

Maize or Field Corn

Corn is susceptible to Mn deficiency, and regions where it can be found are similar to those of other sensitive crops.

Sunflower

There are few references to Mn deficiency in the literature. Sunflower, however, is very tolerant to Mn toxicity, which can be a problem in acid soils (Blamey et al., 1985).

Canola

There have been no reports of Mn deficiency in canola in Canada or the United States. Australia reports that Mn deficiency is possible in high carbonate soils and sandy soils (Hocking et al., 1999).

Sugarbeet

There are no reports of Mn deficiency in sugarbeet in the United States. However, Mn deficiency can be a serious problem in the United Kingdom and it has been controlled using foliar Mn application (Last and Bean, 1991).

Potato

Potato is classified as being susceptible to Mn deficiency in Michigan, Wisconsin, and generally in soils high in organic matter or that tend to be Mn deficient in other sensitive crops. Idaho recommendations for Mn application are based on soil test results (Stark et al., 2004).

Flax

There are few references for flax response to Mn in the literature.

Boron

In the north–central United States, the hot-water B extraction and analysis is useful for identifying B deficiency susceptibility in Wisconsin for alfalfa (*Medicago sativa* L.), but the test is not diagnostic for possible B deficiency in other crops. Boron deficiency is nearly always associated with deep, low-organic-matter sandy soils. Physiological B deficiency symptoms are also possible under very dry and very humid conditions, where B transport within the xylem stream of the plants are very low as a result of limited water movement through the plant. Soybean is categorized as having low susceptibility to B deficiency (Martens and Westermann, 1991). There have been scattered reports of B deficiency in parts of the world such as Thailand (B. Rerkasem et al., 1993) and northeast Arkansas (Ross et al., 2006). Lentil can be susceptible to B deficiency. In Nepal, a program to introduce new lentil germplasm into the country found that 82% of the exotic germplasms tested were B deficient in Nepal soil conditions, whereas native lentil was not deficient and did not respond to B fertilization. Cultivars from Syria were particularly susceptible to B deficiency (Srivastava et al., 2000).

Outside of greenhouse studies, there is little in the literature to support general B application to field pea. Some Canadian publications indicate that both field pea and canola might respond to B if soil test levels are low, but in the case of canola, field trials, both replicated and strip trials, have not shown a yield response (Karamanos et al., 2003). The study also indicated that current Canadian soil test calibration for canola, at least, is incorrect, and critical values from the hot-water B extraction are too high.

There are several reports of B deficiency and correction with B fertilization for wheat in the literature from Pakistan (Tahir et al., 2009) and Bangladesh (Halder et al., 2007). Barley is similar to wheat in its susceptibility to B deficiency, and in both wheat and barley, there is considerable genetic differences in susceptibility between cultivars (Jamjod and Rekasem, 1999).

There are few references in the literature to yield increases in oat from B fertilization. Corn is susceptible to B deficiency, but its distribution in the world is confined mostly to sandy, low-organic-matter soils in regions with sufficient rainfall that would result in leaching, such as Georgia, US (Touchton and Boswell, 1975). Sunflower is particularly susceptible to B deficiency and deficiencies are reported in several places in the world including Australia (Dear and Weir, 2004), South Africa (Blamey et al., 1978), and Russia (Dranichnikova, 1978). Sugarbeet is very susceptible to B deficiency; however, in major sugarbeet growing areas of the United States, including California, Minnesota, Colorado, Wyoming, Montana, and North Dakota, soils have sufficient B that yield and quality response to B fertilization is not seen. Reports of deficiency come from The Netherlands (Smilde, 1970) and Germany (Brandenburg, 1931).

Greenhouse studies in Wisconsin have shown that B deficiency in several Wisconsin soils was possible, that the hot-water B extraction was a good predictor of response in those soils, and that fertilizer B corrected the deficiency in terms of yield and heart rot incidence and severity (Berger and Truog, 1944). Recently in North Dakota and Minnesota trials, sugarbeet studies on sandy, low-organic-matter soils responded to preplant B and foliar B applications (Overstreet et al., 2007). Traditional sugarbeet growing areas in the Red River Valley are high-organic-matter (3–7%) silt loam to clay soils that have not responded to B application in previous studies. As a result of root rot disease incursion in the area, sugarbeet production has moved into more sandy soils at the edges of the Red River Valley. These sandy soils might benefit from the application of B, supported by evidence from a combination of soil test B and understanding soil texture and organic matter characteristics. Flax is one of the most tolerant crops to low soil B (Berger, 1949). Boron deficiency in flax has not been recorded in North Dakota. Evidence for B deficiency in flax is rare in the literature outside of nutrient solution studies.

Molybdenum

Molybdenum deficiencies are nearly always confined to areas with pH $\leq$ 5.5 (Weir, 2004.) Molybdenum functions in plants are associated with N metabolism, so legumes of all kinds are particularly vulnerable to low Mo availability and low-pH-induced deficiency. Deficiencies in legumes in Wisconsin in very sandy soils derived from quartz minerals and low in pH have been recorded (Schulte, 1992). Molybdenum is a greater problem to many crops, especially legumes, in tropical soils with high levels of sesquioxides and low pH (Pasricha et al., 1997).

References

Abi-Ghanem, R., L. Carpenter-Boggs, J.L. Smith, and G.J. Vandemark. 2012. Nitrogen fixation by US and Middle Eastern chickpeas with commercial and wild Middle Eastern inocula. ISRN Soil Sci. 2012:981842. doi:10.5402/2012/981842

Ahlawat, I.P.S. 1990. Diagnosis and alleviation of mineral nutrient constraints in chickpea. In: H.A. van Rheenen and M.C. Saxena, editors, Chickpea in the nineties. International Crops Research Institute for the Semi Arid Tropics, Patancheru, Andhra Pradesh. p. 93–99.

Alcantara, E., and M.D. de la Guardia. 1987. Differential response of sunflower genotypes to iron deficiency. In: H.W. Gubelman and B.C. Loughman, editors, Genetic aspects of plant mineral nutrition (Developments in plant and soil science). Chap. 27. Martinus Nijhoff, Publishers, Boston, MA. p. 457–462.

Allaby, R.G., G. Peterson, D.A. Merriweather, and Y.B. Fu. 2005. Evidence of the domestication history of flax (Linum usitatissimum L.) from genetic diversity of the sad2 locus. Theor. Appl. Genet. 112:58–65. doi:10.1007/s00122-005-0103-3

Alley, M.M., C.I. Rich, G.W. Hawkins, and D.C. Martens. 1978. Correction of Mn deficiency of soybeans. Agron. J. 70:35–38. doi:10.2134/agronj1978.00021962007000010009x

Asadi Rahmani, H., L.A. Rasanen, M. Afshari, and K. Lindstrom. 2011. Genetic diversity and identification of common bean (Phaseolus vulgaris L.) nodulating Rhizobia isolated from soils of Iran. Appl. Soil Ecol. 48:287–293. doi:10.1016/j.apsoil.2011.04.010

Badr, A., K. Muller, R. Schafer-Pregl, H. El Rabey, S. Effgen, H.H. Ibrahim, C. Pozzi, W. Rohde, and F. Salamini. 2000. On the origin and domestication history of barley (Hordeum vulgare). Mol. Biol. Evol. 17:499–510. doi:10.1093/oxfordjournals.molbev.a026330

Bakry, A.B., T.A. Elewa, and A.M. Ali. 2012. Effect of Fe foliar application on yield and quality traits of some flax varieties grown under newly reclaimed sandy soil. Aust. J. Basic Appl. Sci. 6:532–536.

Barbagelata, P.A. 2006. Evaluation of potassium soil tests and methods for mapping soil fertility properties in Iowa corn and soybean fields. Ph.D. thesis, Iowa State Univ., Ames.

Berg, W.A., M.E. Hodges, and E.G. Krenzer. 1993. Iron deficiency in wheat grown on the Southern plains. J. Plant Nutr. 16:1241–1248. doi:10.1080/01904169309364609

Berger, K.C. 1949. Boron in soils and crops. Adv. Agron. 1:321–351. doi:10.1016/S0065-2113(08)60752-X

Berger, K.C., and E. Truog. 1944. Boron deficiency in beets as correlated with yields and available boron. In: B.B. Morgan, editor, Trans. of the Wisconsin Academy of Sciences, Arts and Letters. Vol. XXXVI. p. 421-425.

Berti, M., S. Fischer, R. Wilckens, and F. Hevia. 2009. Flaxseed response to N, P, and K fertilization in South Central Chile. Chilean J. Agric. Res. 69:145–153. doi:10.4067/S0718-58392009000200003

Biddle, A.J., and N.D. Cattlin. 2007. Pests, diseases and disorders of peas and beans. Manson Publishers Ltd., London.

Binford, G.G., G.W. Herbert, and J.M. Blumenthal. 2000. Sugarbeets. In: R.B. Ferguson, editor, Nutrient management for agronomic crops in Nebraska. Univ. of Nebraska Ext. Circ. EC 155. Univ. of Nebraska Ext. Serv., Lincoln, NE. p. 127–130.

Bitocchi, E., L. Nanni, E. Bellucci, M. Rossi, A. Giardini, P.S. Zeuli, G. Logozzo, J. Stougaard, P. McClean, G. Attene, and R. Papa. 2012. Mesoamerican origin of the common bean (*Phaseolus vulgaris* L.) is revealed by sequence data. Proc. Natl. Acad. Sci. USA 109:E788–E796. doi:10.1073/pnas.1108973109

Blackman, B.B., M. Scascitelli, N.C. Kane, H.H. Luton, D.A. Rasmussen, R.A. Bye, D.L. Lentz, and L.H. Rieseberg. 2011. Sunflower domestication alleles support single domestication center in eastern North America. Proc. Natl. Acad. Sci. USA 108:14360–14365. doi:10.1073/pnas.1104853108

Blamey, F.P.C., C.J. Asher, and D.G. Edwards. 1985. Sunflower response to acid soil factors. In: J.J. Yates, editor, Crop and pasture production-science and practice. 3rd Australian Agronomy Conf., 1985. Univ. of Tasmania, Hobart, Tasmania.

Blamey, F.P.C., D. Mould, and K. Nathanson. 1978. Relationships between B deficiency symptoms in sunflower and the B Ca/B status of plant tissues. Agron. J. 70:376–380. doi:10.2134/agronj1978.000219 62007000030004x

Bliss, F.A. 1993. Breeding common bean for improved biological nitrogen fixation. Plant Soil 152:71–79. doi:10.1007/BF00016334

Bolland, M.D.A., K.H.M. Siddique, S.P. Loss, and M.J. Baker. 1999. Comparing responses of grain legumes, wheat and canola to applications of superphosphate. Nutr. Cycl. Agroecosyst. 53:157–175. doi:10.1023/A:1009798506480

Borges, R., and A.P. Mallarino. 2003. Broadcast and deep-band placement of phosphorus and potassium for soybean managed with ridge tillage. Soil Sci. Soc. Am. J. 67:1920–1927. doi:10.2136/sssaj2003.1920

Boring, T., and K. Thelen. 2010. Manganese management of soybeans on chronically Mn deficient soils. Abstract, ASA Abstracts. Madison, WI.

Brandenburg, E. 1931. Die Herz-Trochenfaule der Ruben als Bormangel-Erscheinug (The finding of heart/dry rot of beet as boron deficiency). Phytopathol. Z. 3:499–519.

Bremer, E., C. van Kessel, and R. Karamanos. 1989. Inoculant, phosphorus and nitrogen responses of lentil. Can. J. Plant Sci. 69:691–701. doi:10.4141/cjps89-085

Brennan, R.F., and M.D.A. Bolland. 2007. Comparing the potassium requirements of canola and wheat. Aust. J. Agric. Res. 58:359–366. doi:10.1071/AR06244

Brouder, S.M., A.S. Bongen, K.J. Eck, and S.E. Hawkins. 2003. Manganese deficiencies in Indiana soils. Purdue Univ. Coop. Ext. Serv. Agronomy Guide AY-276-W. Lafayette, IN.

Bu, H. 2014. Yield and quality prediction using satellite passive imagery and ground-based active optical sensors in sugarbeet, spring wheat, corn and sunflower. M.S. thesis. North Dakota State Univ., Soil Science, Fargo.

Buah, S.S.J., T.A. Polito, and R. Killorn. 2000. No-tillage soybean response to banded and broadcast and direct and residual phosphorus and potassium applications. Agron. J. 92:657–662. doi:10.2134/agronj2000.924657x

Butchee, K.S., J. May, and B. Arnall. 2011. Sensor based nitrogen management reduced nitrogen and maintained yield. Crop Management 10. doi:10.1094/CM-2011-0725-01-RS

Cakmak, I. 2008. Zinc deficiency in wheat in Turkey. In: Alloway, B.J, editor, Micronutrient deficiencies in global crop production. Springer Science+Business Media B.V., London. p. 181–200.

Cattanach, N.R., and L. Overstreet. 2006. Effect of potassium fertilizer on sugar production. 2006 Sugarbeet Research and Ext. Reports. Sugarbeet Research and Education Board of Minnesota and North Dakota. North Dakota State Univ. Ext. Serv., Fargo, ND.

Christenson, D.R., and A.P. Draycott. 2006. Nutrition: Phosphorus, sulphur, potassium, sodium, calcium, magnesium and micronutrients-liming and nutrient deficiencies. In: A.P. Draycott, editor, Sugarbeet. Blackwell Publ., Ltd., Oxford, UK. p. 185–200.

Dampney, P., S. Wale, and A. Sinclair. 2011. Potash requirements of potato. Potato Council Report No. 2011/4. Agric. & Hort. Dev. Board, Warwickshire, UK.

Darby, H., E. Cummings, R. Madden, and S. Monahan. 2013. 2012 Sunflower population and nitrogen rate trial. Univ. of Vermont Ext. Publ., Burlington, VT.

Davis, J.D., and M.A. Brick. 2009. Fertilizing dry beans. Colorado State Univ. Ext. Fact Sheet No. 0.539. Colorado State Univ. Ext. Serv., Fort Collins, CO.

Davis, J.D., R.D. Davidson, and S.Y.C. Essah. 2014. Fertilizing potatoes. Colorado State Univ. Ext. Fact Sheet No. 0.541. Colorado State Univ. Ext., Fort Collins, CO.

Dear, B.S., and R.G. Weir. 2004. Boron deficiency in pastures and field crops. AgFact P1.AC.1 2nd ed. NSW Agriculture, Orange, New South Wales.

Deibert, E.J., and R.A. Utter. 2004. Field pea growth and nutrient uptake: Response to tillage systems and nitrogen fertilizer applications. Commun. Soil Sci. Plant Anal. 35:1141–1165. doi:10.1081/CSS-120030595

Dellinger, A.E., J.P. Schmidt, and D.B. Beegle. 2008. Developing nitrogen fertilizer recommendations for corn using an active sensor. Agron. J. 100:1546–1552. doi:10.2134/agronj2007.0386

Denison, R.F., and E.T. Kiers. 2011. Life histories of symbiotic rhizobia and mycorrhizal fungi. Curr. Biol. 21:R775–R785. doi:10.1016/j.cub.2011.06.018

DeVuyst, E.A., and A.D. Halvorson. 2004. Economics of annual cropping versus crop-fallow in the northern Great Plains as influenced by tillage and nitrogen. Agron. J. 96:148–153. doi:10.2134/agronj2004.0148

Dodd, J.R., and A.P. Mallarino. 2005. Soil-test phosphorus and crop grain yield responses to long-term phosphorus fertilization for corn-soybean rotations. Soil Sci. Soc. Am. J. 69:1118–1128. doi:10.2136/sssaj2004.0279

Doxtator, C.W., and F.R. Calton. 1950. Sodium and potassium content of sugarbeet varieties in some western beet growing areas. Proc. Am. Soc. of Sugarbeet Technologists 6:144–151.

Dranichnikova, T.D. 1978. Effect of trace elements on physiological processes and productivity of sunflower. Soils Fert. 42:41–75.

Ekin, Z. 2010. Performance of phosphate solubilizing bacteria for improving growth and yield of sunflower (*Helianthus annuus* L.) in the presence of phosphorus fertilizer. Afr. J. Biotechnol. 9:3794–3800.

Ellis, J.R. 1998. Post flood syndrome and vesicular–arbuscular mycorrhizal fungi. J. Prod. Agric. 11:200–204. doi:10.2134/jpa1998.0200

Ellsworth, J.W., V.D. Jolley, D.S. Nuand, and A.D. Blaylock. 1997. Screening for resistance to iron deficiency chlorosis in dry bean using iron reduction capacity. J. Plant Nutr. 20:1489–1502. doi:10.1080/01904169709365351

Erman, M., E. Ari, Y. Togay, and F. Cig. 2009. Response of field pea (*Pisum sativum* sp. *Arvense* L.) to *Rhizobium* inoculation and nitrogen application in eastern Anotolia. J. Anim. Vet. Adv. 8:612–616.

Esmail, A.O., H.S. Yasin, and B.J. Mahmood. 2014. Effect of levels of phosphorus and iron on growth, yield and quality of flax. IOSR J. Agric. Vet. Sci. 7:7–11.

Fanning, C., and J. Goos. 1992. Phosphorus: Impact on small grain plant development. Better Crops with Plant Food. Potash & Phosphate Institute, Norcross, GA.

Farmaha, B.S., F.G. Fernández, and E.D. Nafziger. 2011. No-till and strip-till soybean production with surface and subsurface phosphorus and potassium fertilization. Agron. J. 103:1862–1869. doi:10.2134/agronj2011.0149

Farrer, D.C., R. Weisz, R. Heiniger, and J.P. Murphy. 2006. Minimizing protein variability in soft red winter wheat: Impact of nitrogen application timing and rate. Agron. J. 98:1137–1145. doi:10.2134/agronj2006.0039

Fernández, F.G., and R.G. Hoeft. 2009. Managing soil pH and crop nutrients In: Illinois Agronomy Handbook. 24th ed. Chap 8. Univ. of Illinois Ext. Serv., Urbana, IL. p. 91–112.

Fernández, F.G., and D. Schaefer. 2012. Assessment of soil phosphorus and potassium following real time kinematic-guided broadcast and deep-band placement in strip-till and no-till. Soil Sci. Soc. Am. J. 76:1090–1099. doi:10.2136/sssaj2011.0352

Fischer, H.E. 1989. Origin of the 'Weisse Schlesische Rube' (white Silesian beet) and resynthesis of sugarbeet. Euphytica 41:75–80. doi:10.1007/BF00022414

Fixen, P.E. 2002. Soil test levels in North America. Better Crops Plant Food 86:12–15.

Fixen, P.E., J.R. Gerwing, and B.G. Farber. 1984. Phosphorus requirements of corn and soybeans following fallow. In: Southeast South Dakota Experiment Farm Progress Reports in Plant Science 84-23. South Dakota State Univ. Agric. Exp. Station, Brookings, SD. p. 72–80.

Flaten, D. 2013. Soluble P losses in spring snow melt runoff in the Northern Great Plains. In: 2013 Great Plains Soil Fertility Conf. Vol. 15. Denver, CO. 4–5 Mar. 2013. Int. Plant Nutrition Inst., Peachtree Corners, GA. p. 47–52.

Flor, H.H. 1943. Chlorotic dieback of flax grown on calcareous soils. J. Am. Soc. Agron. 35:259–269. doi:10.2134/agronj1943.00021962003500040001x

Ford Lloyd, B.V. 1995. Sugarbeet, and other cultivated beets. In: J. Smartt and N.W. Simmonds, editors, Evolution of crop plants. Longman Scientific & Technical, Essex, UK. p. 35–40.

Frank, K., D. Beegle, and J. Denning. 1998. Phosphorus. In: J.R. Brown, editor, Recommended chemical soil test procedures for the North Central Region. North Central Regional Res. Publ. No. 221 (Revised). Missouri Agric. Exp. Station, Columbia, MO. p. 21–30.

Franzen, D.W. 2003. Fertilizing sugarbeet. NDSU Ext. Circ. SF-714 (Revised). North Dakota State Univ. Ext. Serv., Fargo, ND.

Franzen, D.W. 2004a. Delineating nitrogen management zones in a sugarbeet rotation using remote sensing: A review. J. Sugar Beet Res. 41:47–60. doi:10.5274/jsbr.41.1.47

Franzen, D.W. 2004b. Fertilizing flax. NDSU Ext. Circ. 717 (Revised). North Dakota State Univ. Ext. Serv., Fargo, ND.

Franzen, D.W. 2004c. Flax fertility recommendation changes in North Dakota. In: Proc. of the 2004 North Central Extension-Industry Soil Fertility Conf., Des Moines, IA. Nov. 2004. Potash and Phosphate Institute, Brookings, SD. p. 144–150.

Franzen, D.W. 2006. Fertilizing pinto, navy, and other dry edible beans. NDSU Ext. Circ. SF-720 (Revised). North Dakota State Univ. Ext. Serv., Fargo, ND.

Franzen, D.W. 2009. Fertilizing hard red spring wheat and durum. NDSU Ext. Circ. SF-712. North Dakota State Univ. Ext. Serv., Fargo, ND.

Franzen, D.W. 2010. North Dakota fertilizer recommendation tables and equations. NDSU Ext. Circ. SF-882 (revised). North Dakota State Univ. Ext. Serv., Fargo, ND.

Franzen, D.W. 2011. Variability of soil test potassium in space and time. In: Proc. of the 41st North Central Extension-Industry Soil Fertility Conf. Vol. 27. Des Moines, IA. 16–17 Nov. International Plant Nutrition Institute, Brookings, SD. p. 74–82.

Franzen, D.W. 2013. Soybean soil fertility. NDSU Ext. Circ. SF-1164. North Dakota State Univ. Ext. Serv., Fargo, ND

Franzen, D.W. 2014. Soil fertility recommendations for corn. NDSU Ext. Circ. SF722 (Revised). North Dakota State Univ. Ext. Serv., Fargo, ND.

Franzen, D.W. 2016. Fertilizing sunflower. NDSU Ext. Circ. SF-713 (Revised). NDSU Ext. Serv., Fargo, ND.

Franzen, D.W., and R.J. Goos. 2007. Fertilizing malting and feed barley. NDSU Ext. Circ. SF-723 (Revised). North Dakota State Univ. Ext. Serv., Fargo, ND.

Franzen, D.W., and C.A. Grant. 2008. Sulfur response based on crop, source, and landscape position. In: J. Jez, editor, Sulfur: Missing link between soils, crops and nutrition. Agronomy Monograph No. 50. ASA, CSSA, and SSSA, Madison, WI. p. 105–116.

Franzen, D.W., and J. Lukach. 2007. Fertilizing canola and mustard. NDSU Ext. Circ. SF-1122 (Revised). North Dakota State Univ. Ext. Serv., Fargo, ND.

Franzen, D.W., M.V. McMullen, and D.S. Mosset. 2008. Spring wheat and durum yield and disease responses to copper fertilization of mineral soils. Agron. J. 100:371–375. doi:10.2134/agronjl2007.0200

Franzen, D.W., T. Nanna, and W.A. Norvell. 2006. A survey of soil attributes in North Dakota by landscape position. Agron. J. 98:1015–1022. doi:10.2134/agronj2005.0283

Franzen, D.W., and J.L. Richardson. 2000. Soil factors affecting iron chlorosis in the Red River Valley of North Dakota and Minnesota. J. Plant Nutr. 23:67–78. doi:10.1080/01904160009381998

Franzen, D.W., L.K. Sharma, and H. Bu. 2014. Active optical sensor algorithms for corn yield prediction and a corn side-dress nitrogen rate aid. NDSU Ext. Circ. SF-1176 (5). North Dakota State Univ. Ext. Serv., Fargo, ND.

French, R., and P. White. 2005. Soil and environmental factors affecting pulse adaptation in Western Australia. In: P. White, M. Seymour, P. Burgess and M. Harries, editors, Producing pulses in the Southern Agric. Region. Bulletin 4645. GRDC Department of Agriculture, Government of Western Australia. p. 1–18.

Fu, Y.B. 2011. Genetic evidence for early flax domestication with capsular dehiscence. Genet. Resour. Crop Evol. 58:1119–1128. doi:10.1007/s10722-010-9650-9

Gallagher, P.H., and T. Walsh. 1943. The susceptibility of cereal varieties to manganese deficiency. J. Agric. Sci. 33:197–203. doi:10.1017/S0021859600007152

Gan, Y., G.W. Clayton, G. Lafond, A. Johnston, F. Walley, and B.G. McConkey. 2004. Effect of 'starter' N and P on nodulation and seed yield in field pea, lentil and chickpea in semiarid Canadian Prairies. Univ. of Saskatchewan Soils and Crops Conf. Proc. http://www.usask.ca/soilsncrops/conference-proceedings/previous_years/Files/2004/2004docs/062.pdf (accessed 28 Sept. 2014).

Gan, Y., A.M. Johnston, J.D. Knight, C. McDonaold, and C. Stevenson. 2010. Nitrogen dynamics of chickpea: Effects of cultivar choice, N fertilization, Rhizobium inoculation, and cropping systems. Can. J. Plant Sci. 90:655–666. doi:10.4141/CJPS10019

Garcia, R.L., and J.J. Hanway. 1976. Foliar fertilization of soybeans during the seed-filling period. Agron. J. 68:653–657. doi:10.2134/agronj1976.00021962006800040030x

Gelderman, R., J. Gerwing, and A. Bly. 2002. Influence of potassium (K), rate, placement and hybrid on K deficiency in corn. Soil/Water Res. Soil PR 02-10. South Dakota State Univ., Brookings, SD.

Geleta, S., D.D. Baltensperger, G.D. Binford, and J.F. Miller. 1997. Sunflower response to nitrogen and phosphorus in wheat-fallow cropping systems. J. Prod. Agric. 10:466–472. doi:10.2134/jpa1997.0466

Gerwing, J., and R. Gelderman. 2005. South Dakota fertilizer recommendation guide. South Dakota State Univ. Ext. Serv. Circ. EC750, Brookings, SD.

Glasscock, H.H. 1941. Varietal susceptibility of peas to marsh spot. Ann. Appl. Biol. 28:316–324. doi:10.1111/j.1744-7348.1941.tb07564.x

Godsey, C.B., J.P. Schmidt, A.J. Schlegel, R.K. Taylor, C.R. Thompson, and R.J. Gehl. 2003. Correcting Fe deficiency in corn with seed row-applied Fe sulfate. Agron. J. 95:160–166. doi:10.2134/agronj2003.0160

Government of Western Australia. 2015. Diagnosing manganese deficiency in wheat. Govt. Western Australia, Dep. Agric. and Food, Perth. https://agric.wa.gov.au/n/1993 (accessed 24 Sept. 2014).

Graham, R.D., and E.K.S. Nambiar. 1981. Advances in research on copper deficiency in cereals. Australian Agric. Res. 32:1009–1037.

Grove, J.H., R.C. Ward, and R.R. Weil. 2007. Nutrient stratification in no-till soils. Leading Edge. J. No-Till Agric. 6:374–381.

Grunes, D.L., L.C. Boawn, C.W. Carlson, and F.G. Viets. 1961. Zinc deficiency of corn and potatoes as related to soil and plant analyses. Agron. J. 53:68–71. doi:10.2134/agronj1961.00021962005300020002x

Gupta, S.K., and A. Pratap. 2007. History, origin and evolution. Adv. Bot. Res. 45:1–20. doi:10.1016/S0065-2296(07)45001-7

Hairston, J.E., W.F. Jones, P.K. McConnaughey, L.K. Marshall, and K.B. Gill. 1990. Tillage and fertilizer effects on soybean growth and yield on three Mississippi soils. J. Prod. Agric. 3:317–325. doi:10.2134/jpa1990.0317

Halder, N.K., M.A. Hossain, M.A. Siddiky, N. Nasreen, and M.H. Ullah. 2007. Response of wheat varieties to boron supplication in calcareous brown floodplain soil at southern region of Bangladesh. J. Agron. 6:21–24. doi:10.3923/ja.2007.21.24

Halley, S., and E.J. Deibert. 1996. Canola response to sulfur fertilizer applications under different tillage and landscape position. 1996 Annual Rep. to USDA–CSREES Special Programs, Northern Region Canola and North Dakota Oilseed Council. North Dakota State Univ., Fargo, ND.

Haq, M.U., and A.P. Mallarino. 1998. Foliar fertilization of soybean at early vegetative stages. Agron. J. 90:763–769. doi:10.2134/agronj1998.00021962009000060008x

Harmer, P.M. 1946. Studies on the effect of copper sulfate applied to organic coil on the yield and quality of several crops. Soil Sci. Soc. Am. Proc. 10:284–294. doi:10.2136/sssaj1946.03615995001000C00050x

Helms, T.C., R.A. Scott, W.T. Schapaugh, R.J. Goos, D.W. Franzen, and A.J. Schlegel. 2010. Soybean iron-deficiency chlorosis tolerance and yield decrease on calcareous soils. Agron. J. 102:492–498. doi:10.2134/agronj2009.0317

Hendrickson, P. 2007. Corn response to starter fertilizers. Carrington Res. & Ext. Center 2007 Annual Rep. North Dakota State Univ. Agric. Exp. Station, Carrington Res. and Ext. Center, Carrington, ND.

Hendrickson, P., B. Henson, and J. Lukach. 2008. Effect of phosphorus placement and rate on canola. 2008 Carrington Res. & Ext. Center Annual Rep. North Dakota State Univ. Agric. Exp. Station, Carrington Res. and Ext. Center, Carrington, ND.

Henson, R. 2004. Field pea response to nitrogen fertilizer. 2004 Carrington Res. & Ext. Center Res. Rep., North Dakota State Univ. Exp. Station. Carrington, ND.

Hergert, G.W. 2010. Sugarbeet fertilization. Sugar Tech: An International Journal of Sugar Crops and Related Industries 12:256–266.

Hergert, G.W., P.T. Nordquist, J.L. Petersen, and B.A. Skates. 1996. Fertilizer and crop management practices for improving maize yields on high pH soils. J. Plant Nutr. 19:1223–1233. doi:10.1080/01904169609365193

Hergert, G.W., and J.A. Schild. 2013. Fertilizer management for dry edible beans. Univ. of Nebraska Ext. Circ. G1713 (Revised). Univ. of Nebraska Ext. Serv., Lincoln, NE.

Hocking, P., R. Norton, and A. Good. 1999. Crop Nutrition. In: 10th International Rapeseed Congress, Canberra, Australia.

Holland, K.H., and J.S. Schepers. 2013. Use of a virtual-reference concept to interpret active crop canopy sensor data. Precis. Agric. 14:71–85. doi:10.1007/s11119-012-9301-6

Hoyman, W.G. 1949. The effect of zinc-containing dusts and sprays on the yield of potatoes. Am. Potato J. 26:256–263. doi:10.1007/BF02883860

Hoyt, P.B., and G.G.S. Myovella. 1979. Correction of severe manganese deficiency in wheat with chemical fertilizers. Plant Soil 52:437–444. doi:10.1007/BF02185586

Huber, D.M., J.D. Leuck, W.C. Smith, and E.P. Christmas. 2004. Induced manganese deficiency in GM soybeans. In: Proc. of the 2004 North Central Extension-Industry Soil Fertility Conf., Des Moines, IA. 17–18 Nov. International Plant Nutrition Institute, Brookings, SD. p. 80–83.

Husted, S., K.H. Laursen, C.A. Hebbern, S.B. Schmidt, P. Pedas, A. Haldrup, and P.E. Jensen. 2009. Manganese deficiency leads to genotype-specific changes in fluorescence induction kinetics and state transitions. Plant Physiol. 150:825–833. doi:10.1104/pp.108.134601

Iltis, H.H., and J.F. Doebley. 1980. Taxonomy of *Zea* (Gramineae). II. Subspecific categories in the *Zea mays* complex and generic synopsis. Am. J. Bot. 67:994–1004. doi:10.2307/2442442

Inskeep, W.P. and P.R. Bloom. 1986. Effects of soil moisture on soil pCO_2, soil solution bicarbonate, and iron chlorosis in soybean. Soil Sci. Soc. Am. J. 50:946–952.

Jamjod, S., and B. Rekasem. 1999. Genotypic variation in response of barley to boron deficiency. Plant Soil 215:65–72. doi:10.1023/A:1004736610871

Johnson, R.G., and M.B. Ali. 1982. Economics of wheat-fallow cropping systems in Western North Dakota. West. J. Agric. Econ. 7:67–78.

Kaan, D.A., D.M. O'Brien, P.A. Burgener, G.A. Peterson, and D.G. Westfall. 1982. An economic evaluation of alternative crop rotations compared to wheat-fallow in northeastern Colorado. Colorado Tech. Bull. TB02-1. Colorado State Univ., Fort Collins, CO.

Kaiser, D.E., and K.I. Kim. 2013. Soybean response to sulfur fertilizer applied as a broadcast or starter using replicated strip trials. Agron. J. 105:1189–1198. doi:10.2134/agronj2013.0023

Kaiser, D.E., A.P. Mallarino, and M. Bermudez. 2005. Corn grain yield, early growth, and early nutrient uptake as affected by broadcast and in-furrow starter fertilization. Agron. J. 97:620–626. doi:10.2134/agronj2005.0620

Kandel, H. 2012. Soybean nodulation (6/21/2012). Crop and Pest Report. North Dakota State Univ., Fargo. https://www.ag.ndsu.edu/cpr/plant-science/soybean-nodulation-6-21-12 (accessed 27 Sept. 2014).

Kansas State University. 2015. Fertilizer recommendations. KSU Dept. of Agronomy, Manhattan, KS. http://www.agronomy.ksu.edu/soiltesting/p.aspx?tabid=32 (accessed 25 Sept. 2014).

Kaplan, L., and T.F. Lynch. 1999. *Phaseolus* (Fabaceae) in Archaeology: AMS radiocarbon dates and their significance for Pre-Columbian agriculture. Econ. Bot. 53:261–272. doi:10.1007/BF02866636

Karamanos, R., and N. Flore. 2000. Starter potassium for wheat and barley on high potassium soils. Better Crops Plant Food 84:14–15.

Karamanos, R.E., N.A. Flore, and J.T. Harapiak. 2003. Response of field peas to phosphate fertilization. Can. J. Plant Sci. 83:283–289. doi:10.4141/P02-110

Karamanos, R., N.A. Flore, and J.T. Harapiak. 2013. Application of seed-row potash to spring wheat grown on soils with high available potassium levels. Can. J. Plant Sci. 93:271–277. doi:10.4141/cjps2012-148

Karamanos, R.E., T.B. Goh, and D.N. Flaten. 2007. Nitrogen and sulphur fertilizer management for growing canola on sulphur sufficient soils. Can. J. Plant Sci. 87:201–210. doi:10.4141/P06-133

Karamanos, R.E., T.B. Goh, and T.A. Stonehouse. 2003. Canola response to boron in Canadian prairie soils. Can. J. Plant Sci. 83:249–259. doi:10.4141/P02-095

Kelling, K.A., E. Panique, P.E. Speth, and W.R. Stevenson. 2002. Effect of potassium rate, source and application timing on potato yield and quality. Proc. of the Idaho Potato Conf., 23 Jan. 2002.

Keyser, H.H., and F. Li. 1992. Potential for increasing biological nitrogen fixation in soybean. Plant and Soil 141:119–135.

Khan, H.R., G.K. McDonald, and Z. Rengel. 2000. Response of chickpea genotypes to zinc fertilization under field conditions in south

Australia and Pakistan. J. Plant Nutr. 23:1517–1531. doi:10.1080/01904160009382119

Khan, M.A., J. Din, S. Nasreen, M.Y. Khan, S.U. Khan, and A.R. Gurmani. 2009. Response of sunflower to different levels of zinc and iron under irrigated conditions. Sarhad J. Agric. 25:159–163.

Khattak, S.G., D.F. Khan, S.H. Shah, M.S. Madani, and T. Khan. 2006. Role of Rhizobial inoculation in the production of chickpea crop. Soil Environ. (Faisalabad, Pak.) 25:143–145.

King, P.M., and A.M. Alston. 1975. Diagnosis of trace element deficiencies in wheat on Eyre Peninsula, South Australia. In: J.D. Nicholas and A.R. Egan, editors, Trace elements in soil–plant–animal systems. Academic Press, NY. p. 339–352.

Kiple, K.F., and K.C. Ornelas. 2000. Cambridge world history of food. Cambridge Univ. Press, NY.

Kopittke, P.M., and N.W. Menzies. 2007. A review of the use of the basic cation saturation ratio and the 'ideal' soil. Soil Sci. Soc. Am. J. 71:259–265. doi:10.2136/sssaj2006.0186

Kroetz, M.E., W.H. Schmidt, J.E. Beverlein, and G.L. Ryder. 1977. Correcting Mn deficiency increases soybean yields. Ohio Rep. 62:51–53.

Kruger, G.A., R.E. Karamanos, and J.P. Singh. 1985. The copper fertility of Saskatchewan soils. Can. J. Soil Sci. 65:89–99. doi:10.4141/cjss85-010

Last, P.J., and K.M.R. Bean. 1991. Controlling manganese deficiency in sugarbeet with foliar sprays. J. Agric. Sci. Cambridge 116:351–358. doi:10.1017/S0021859600078163

Leikam, D.F., R.E. Lamond, and D.B. Mengel. 2003. Soil test interpretations and fertilizer recommendations. Kansas State Agric. Exp. Station and Ext. Serv. Circ. MF-2586, Manhattan, KS.

Li, F., Y. Miao, F. Zhang, Z. Cui, R. Li, X. Chen, H. Zhang, J. Schroder, W.R. Raun, and L. Jia. 2009. In-season optical sensing improves nitrogen-use efficiency for winter wheat. Soil Sci. Soc. Am. J. 73:1566–1574. doi:10.2136/sssaj2008.0150

Lindsay, W.L., and W.A. Norvell. 1978. Development of a DTPA soil test for zinc, iron, manganese and copper. Soil Sci. Soc. Am. J. 42:421–428. doi:10.2136/sssaj1978.03615995004200030009x

Lucas, R.E. 1945. The effect of the addition of sulfates of copper, zinc and manganese on the absorption of these elements by plants grown on organic soils. Soil Sci. Soc. Am. Proc. 9:269–274.

Macnack, N., B.K. Chim, B. Amedy, and B. Arnall. 2014. Fertilization based on sufficiency, build-up and maintenance concept. Oklahoma Cooperative Ext. Serv. Circ. PSS-2266. Oklahoma State Univ., Stillwater, OK.

Mahler, R.L. 2005. Chickpeas. Northern Idaho Fertilizer Guide, CIS 826. Univ. of Idaho, Moscow, ID.

Mallarino, A.P. 2008. Fertilizing crops in the new price age-phosphorus and potassium. In: Proc. 20th Annual Integrated Crop Manag. Conf., Ames, IA. 10–11 Dec. 2008. Iowa State Univ., Ames. p. 261–266.

Mallarino, A.P., R. Borges, and D. Wittry. 2001. Corn and soybean response to potassium fertilization and placement. In: North-Central Extension-Industry Soil Fertility Conf. Proc. Vol. 16. Des Moines, IA. 14–15 Nov. 2001. International Plant Nutrition Institute, Brookings, SD. p. 5–11.

Mallarino, A.P., M.W. Clover, and R.R. Oltmans. 2011. Identification of reasons for high temporal soil-test potassium variation. In: Proc. of the 41st North Central Extension-Industry Soil Fertility Conf. Vol. 27, Des Moines, IA. 16–17 Nov. 2011. International Plant Nutrition Institute, Brookings, SD. p. 65–73.

Martens, D.C., and D.T. Westermann. 1991. Fertilizer applications for correcting micronutrient deficiencies. In: J.J. Mortvedt, F.R. Cox, L.M. Shuman, and R.M. Welch, editors, Micronutrients in agriculture. 2nd ed. SSSA Book Series, No. 4. SSSA, Madison, WI. p. 549–592.

McKenzie, R.H., A.B. Middleton, E.D. Solberg, J. DeMulder, N. Flore, G.W. Clayton, and E. Bremer. 2001. Response of pea to rate and placement of phosphate fertilizer in Alberta. Can. J. Plant Sci. 81:645–649. doi:10.4141/P01-007

Moraghan, J.T. 1979. Manganese toxicity in flax growing on certain calcareous soils low in available iron. Soil Sci. Soc. Am. J. 43:1177–1180. doi:10.2136/sssaj1979.03615995004300060024x

Moraghan, J.T. 1980. Effect of soil temperature on response of flax to phosphorus and fertilizers. Soil Sci. 129:290–296. doi:10.1097/00010694-198005000-00005

Moraghan, J.T. 1982. Zinc deficiency of flax. North Dakota Farm Res. 1982:20–26.

Moraghan, J.T. 1985. Potassium nutrition of sugarbeets. In: R.D. Munson, editor, Potassium in agriculture. ASA, CSSA, and SSSA, Madison, WI. p. 1063–1076.

Mullison, W.R., and E. Mullison. 1942. Growth responses of barley seedlings in relation to potassium and sodium nutrition. Plant Physiol. 17:632–644. doi:10.1104/pp.17.4.632

Mulvaney, D.L., and J.W. Pendleton. 1967. Soybean yields as affected by correction of Mn deficiency in sandy soils. Illinois Res. 9:16.

Neptune, A.M.L., and T. Muraoka. 1977. Effects of foliar fertilization of common bean (*Phaseolus vulgaris* L.) during seed-filling stage. Anais da Escola Superior de Agricultura 'Luiz de Querioz' 34:551–563.

Nielsen, D.C., M.F. Vigil, and J.G. Benjamin. 2011. Evaluating decision rules for dryland rotation crop selection. Field Crops Res. 120:254–261. doi:10.1016/j.fcr.2010.10.011

Nuttall, W.F., and R.G. Button. 1990. The effect of deep banding N and P fertilizer on the yield of canola (*Brassica napus* L.) and spring wheat (*Triticum aestivum* L.). Can. J. Soil Sci. 70:629–639. doi:10.4141/cjss90-066

Nuttall, W.F., A.P. Moulin, and L.J. Townley-Smith. 1992. Yield response of canola to nitrogen, phosphorus, precipitation and temperature. Agron. J. 84:765–768. doi:10.2134/agronj1992.00021962008400050001x

Ontario Ministry of Agriculture. 2014. Spring and winter canola: Fertility management. OMA, Food and Rural Affairs. http://omafra.gov.on.ca/english/crops/pub811/6fertility.htm (accessed 28 Sept. 2014).

Overstreet, L.F., D. Franzen, and N. Cattanach. 2007. 2007 micronutrient studies on sugarbeet. In: 2007 Sugarbeet Res. and Ext. Rep. Sugarbeet Res. and Education Board of Minnesota and North Dakota. North Dakota State Univ. Ext. Serv., Fargo, ND. p. 98–101.

Owen, K.J., T.G. Clewett, and J.P. Thompson. 2010. Pre-cropping with canola decreased *Pratylenchus thornei* populations, arbuscular mycorrhizal fungi, and yield of wheat. Crop Pasture Sci. 61:399–410. doi:10.1071/CP09345

Palta, J.A., A.S. Nandwil, K. Sunita, and N.C. Turner. 2005. Foliar nitrogen applications increase the seed yield and protein content in chickpea (*Cicer arietinum* L.) subject to terminal drought. Aust. J. Agric. Res. 56:105–112. doi:10.1071/AR04118

Pasricha, N.S., V.K. Nayyar, and R. Singh. 1997. Molybdenum in the tropics. In: U.C. Gupta, editor, Molybdenum in agriculture. Cambridge Univ. Press, Cambridge, UK. p. 245–270.

Peleg, Z., T. Fahima, A.B. Korol, S. Abbo, and Y. Saranga. 2011. Genetic analysis of wheat domestication and evolution under domestication. J. Exp. Bot. 62:5051–5061. doi:10.1093/jxb/err206

Penney, D.C., E.D. Solberg, I.R. Evans, and L.J. Peining. 1988. The copper fertility of Alberta soils. In: Great Plains Soil Fertility Workshop Proc., 1988, Kansas State Univ., Manhattan, KS.

Peverill, K.I., L.A. Sparrow, and D.J. Reuter, editors. 1999. Soil analysis: An interpretation manual. CSIRO Publishing, Melbourne, Victoria.

Power, J.F., J.J. Bond, F.M. Sandoval, and W.O. Willis. 1974. Nitrification in Paleocene shale. Science 183:1077–1079. doi:10.1126/science.183.4129.1077

Power, J.F., D.L. Grunes, W.O. Willis, and G.A. Reichman. 1963. Soil temperature and phosphorus effects upon barley growth. Agron. J. 55:389–392. doi:10.2134/agronj1963.00021962005500040028x

Randall, G.W., and R.G. Hoeft. 1988. Placement methods for improved efficiency of P and K fertilizers: A review. J. Prod. Agric. 1:70–79. doi:10.2134/jpa1988.0070

Ranere, A.J., D.R. Piperno, I. Holst, R. Dickau, and J. Irtarte. 2009. The cultural and chronological context of early Holocene maize and squash domestication in the Central Balsas River Valley, Mexico. Proc. Natl. Acad. Sci. USA 106:5014–5018. doi:10.1073/pnas.0812590106

Rasz, G.J. 1980. Method and time of phosphorus application for annual crops. In: Proc. of the 24th Annual Manitoba Soil Science Meeting, 3–4 Dec. 1980. Univ. of Manitoba, Winnipeg, MB. p. 115–130.

Raun, W.R., J.B. Solie, G.V. Johnson, M.L. Stone, R.W. Mullen, K.W. Freeman, W.E. Thomason, and E.V. Lukina. 2002. Agron. J. 94:815–820. doi:10.2134/agronj2002.8150

Raun, W.R., J.B. Solie, M.L. Stone, K.L. Martin, K.W. Freeman, R.W. Mullen, H. Zhang, J.S. Schepers, and G.V. Johnson. 2005. Optical sensor-based algorithms for crop nitrogen fertilization. Commun. Soil Sci. Plant Anal. 36:2759–2781. doi:10.1080/00103620500303988

Redmond, C.E., and H.W. Omodt. 1967. Some till-derived Chernozem soils in Eastern North Dakota: I Morphology, genesis, and classification. Soil Sci. Soc. Am. Proc. 31:89–99. doi:10.2136/sssaj1967.03615995003100010025x

Rehm, G.W. 1986. Response of irrigated soybeans to rate and placement of fertilizer phosphorus. Soil Sci. Soc. Am. J. 50:1227–1230. doi:10.2136/sssaj1986.03615995005000050028x

Rehm, G.W. 1995. Impact of banded potassium for corn and soybean production in a ridge-till planting system. Commun. Soil Sci. Plant Anal. 26:2725–2738. doi:10.1080/00103629509369482

Rehm, G.W., and J.A. Lamb. 2004. Impact of banded potassium on crop yield and soil potassium in ridge-till planting. Soil Sci. Soc. Am. J. 68:629–636. doi:10.2136/sssaj2004.6290

Rengel, Z., and V. Romheld. 2000. Differential tolerance to Fe and Zn deficiencies in wheat germplasm. Euphytica 113:219–225. doi:10.1023/A:1003965007305

Rerkasem, B., R.W. Bell, S. Lodkaew, and J.F. Loneragan. 1993. Boron deficiency in soybean (*Glycine max* L. Merr.), peanut (*Arachis hypogaea* L.) and black gram (*Vigna mungo* L. Hepper): Symptoms in seeds and differences among soybean cultivars in susceptibility to boron deficiency. Plant and Soil 150:289–294. doi:10.1007/BF00013026

Reuter, D. 2007. Trace element disorders in South Australian Agriculture. http://www.pir.sa.gov.au/__data/assets/pdf_file/0011/49619/Trace_Element_disorders_in_SA.pdf (accessed 28 Sept. 2014).

Ritchie, K.B., R.G. Hoeft, E.D. Nafziger, L.C. Gonzini, and J.J. Warren. 1997. Starter fertilizers for minimum-till corn. In: Illinois Fertilizer Conf. Proc. 27–29 Jan. 1997. http://frec.ifca.com/1997/report3/index.htm (accessed 28 Sept. 2014).

Ritchie, K.B., R.G. Hoeft, L.C. Gonzini, and J.J. Warren. 1998. Starter fertilizers for reduced-till corn. Illinois Fertilizer Conf. Proc. 26–28 Jan. 1998. http://frec.ifca.com/1998/report10/index.htm (accessed 28 Sept. 2014).

Rose, I.A., W.L. Felton, and L.W. Banks. 1981. Responses of four soybean varieties to foliar zinc fertilizer. Aust. J. Exp. Agric. 21:236–240. doi:10.1071/EA9810236

Rosen, C.J. 2014. Potato fertilization on irrigated soils. Univ. of Minnesota Ext. Serv., St. Paul, MN. http://www.extension.umn.edu/agriculture/crops/potato-fertilization-on-irrigated-soils/ (accessed 27 Sept. 2014).

Rosen, C.J., K.A. Kelling, J.C. Stark, and G.A. Porter. 2014. Optimizing phosphorus fertilizer management in potato production. Am. J. Potato Res. 91:145–160. doi:10.1007/s12230-014-9371-2

Rosolem, C.A., G.J.M. de Andrade, I.P. Lisboa, and S.M. Zoca. 2010. Manganese uptake and redistribution in soybean as affected by glyphosate.

Rev. Bras. Cienc. Solo 34:1915–1922. doi:10.1590/S0100-06832010000600016

Ross, J.R., N.A. Slaton, K.R. Brye, and R.E. DeLong. 2006. Boron fertilization influences on soybean yield and leaf and seed boron concentrations. Agron. J. 98:198–205. doi:10.2134/agronj2005-0131

Rowland, G., editor. 2014. Growing flax: Production, management & diagnostic guide. 4th ed. Flax Council of Canada, Winnipeg, MB, Canada, and Saskatchewan Flax Development Commission, Saskatoon, Saskatchewan, Canada. http://www.agriculture.gov.sk.ca/Default.aspx?DN=0aa6663b-c240-4594-8889-9f54de340c2b (accessed 27 Sept. 2014).

Ryan, M.H., and J.F. Angus. 2003. Arbuscular mycorrhizae in wheat and field pea crops on a low P soil: Increased Zn-uptake but no increase in P-uptake or yield. Plant Soil 250:225–239. doi:10.1023/A:1022839930134

Sakal, R., R.B. Sinha, and A.P. Singh. 1987. Response of chickpea to iron application in a calcareous soils. Int. Chickpea Newsl. 16:14–16.

Sawyer, J.E., B. Lang, and D.W. Barker. 2012. Sulfur fertilization response in Iowa corn and soybean production. In: 2012 Proc. of the Wisconsin Crop Management Conf. Vol. 51. Madison, WI. p. 39–48.

Sawyer, J., A.P. Mallarino, and M. Al-Kaisi. 2011. Flooded soil syndrome. Flood Recovery for Cropland Fact Sheet. Univ. of Nebraska Ext. Serv., Lincoln, NE, and Iowa State Univ., Ames, IA.

Sawyer, J., and E. Nafziger. 2005. Regional approach to making nitrogen fertilizer rate decisions for corn. In: Proc. of the North Central Extension-Industry Soil Fertility Conf. Des Moines, IA. 16–17 Nov. 2005. Potash and Phosphate Inst., Brookings, SD. p. 16–24.

Sawyer, J., E. Nafziger, G. Randall, L. Bundy, G. Rehm, and B. Joern. 2006. Concepts and rationale for regional nitrogen rate guidelines for corn. PM 2015. Iowa State Univ. Ext., Ames, IA.

Saxena, M.C., and A.C. Sheldrake. 1980. Iron chlorosis in chickpea (Cicer arietinum L.) grown on high pH, calcareous vertisols. Field Crops Res. 3:211–214. doi:10.1016/0378-4290(80)90029-5

Scharf, P., and J. Lory. 2006. Best management practices for nitrogen fertilizer in Missouri. Plant Protection programs, Bull. IPM1027, College of Agriculture, Food and Natural Resources, Univ. of Missouri, Columbia, MO.

Scheiner, J.D., F.H. Gutierrez-Boem, and R.S. Lavado. 2002. Sunflower nitrogen requirement and ^{15}N fertilizer recovery in Western Pampas, Argentina. Eur. J. Agron. 17:73–79. doi:10.1016/S1161-0301(01)00147-2

Schulte, E.E. 1992. Soil and applied molybdenum. Univ. of Wisconsin Ext. Circ. A3555. Madison, WI.

Shaker, A.T., and S.A. Al-Doori. 2012. Response of some sunflower hybrids to zinc foliar spraying and phosphorus fertilizer levels under sandy soils conditions. J. Tikrit Univ. Agric. Sci. 17:174–182.

Sharma, L.K., D.W. Franzen, and H. Bu. 2013. Evaluation of wavelength from ground-based active optical sensors for corn yield prediction in North Dakota. In: Proc. of the 43rd North Central Extension-Industry Soil Fertility Conf. Vol. 29. Des Moines, IA. 20–21 Nov. 2013. International Plant Nutrition Institute, Brookings, SD. p. 117–122.

Shepherd, L.N., J.C. Shickluna, and J.F. Davis. 1959. The sodium–potassium nutrition of sugar beets produced on organic soil. J. Am. Soc. Sugarbeet Technologists 10:603–608.

Sidhu, A.S., S.S. Dhaliwal, J.S. Brar, and J.S. Manchanda. 2012. Amelioration of aggravated manganese deficiency in wheat crop through field demonstrations in light textured soils of Punjab. J. Soils Crops. 12:80–83.

Sims, A.L. 2006. Sugarbeet production on sandy soils: The need for non-traditional nutrients. In: 2006 Sugarbeet Res. and Ext. Rep. Sugarbeet Res. and Education Board of Minnesota and North Dakota. North Dakota State Univ. Ext. Serv., Fargo, ND.

Sims, A.L., and L.J. Smith. 2001. Early growth response of sugarbeet to fertilizer phosphorus in phosphorus deficient soils of the Red River Valley. J. Sugar Beet Res. 38:1–17. doi:10.5274/jsbr.38.1.1

Slaton, N.A., R.E. DeLong, M. Mozaffari, J. Shafer, J. Branson, and T. Richards. 2005. Soybean response to phosphorus and potassium fertilization rate on silt loam soils. Arkansas Agric. Exp. Stn. Res. Ser. 548:63–67.

Smilde, K.W. 1970. Soil Analysis as a Basis for Boron Fertilization of Sugar Beets. Zeitschrift for Pflanzenernahrung and Bodenkunde (J. Plant Soil Sci. Germany) 125:130–143. doi:10.1002/jpln.19701250206

Solari, F., J. Shanahan, R.B. Ferguson, J.S. Schepers, and A.A. Gitelson. 2008. Active sensor reflectance measurements of corn nitrogen status and yield potential. Agron. J. 100:571–579. doi:10.2134/agronj2007.0244

Specht, J.E., A. Bastidas, F. Salvagiotti, T. Setiyono, A. Liska, A. Dobermann, D.T. Walters, and K.G. Cassman. 2006. Soybean yield potential and management practices required to achieve it. Univ. Neb. Annual Report NE-11F. http://soysim.unl.edu/publications/specht_et_al_2006_soybean_yield_potential_and_management.pdf (accessed 14 Dec. 2015).

Srivastava, S.P., T.M.S. Bhandari, C.R. Yadav, M. Joshi, and W. Erskine. 2000. Boron deficiency in lentil: Yield loss and geographic distribution in a germplasm collection. Plant Soil 219:147–151. doi:10.1023/A:1004708700075

Stark, J., D. Westermann, and B. Hopkins. 2004. Nutrient management guidelines for Russet Burbank potatoes. Univ. of Idaho Ext. Bull. 840. Univ. Idaho, Moscow, ID.

Stevens, B., and K. Belden. 2005. Nutrient management guidelines for dry beans of Wyoming. Univ. Wyoming Coop. Ext. Laramie, WY.

Stevens, W.B., and A.O. Mesbah. 2005. Zinc sulfate applied to sugarbeet using broadcast, seed-placed and foliar methods. In: Proc. Western Nutrient Management Conf. Vol. 6, Salt Lake City, UT. 2 Mar. 2005. 6:200–207.

Tahir, M., A. Tanveer, T.H. Shah, N. Fiaz, and A. Wasaya. 2009. Yield response of wheat (*Triticum aestivum* L.) to boron application at different growth stages. Pakistan J. Life Soc. Sci. 7:39–42.

Taylor, R., and P. Sylvester. 2011. Weekly crop update, April 8, 2011. Delaware Ext. Serv. http://agdev.anr.udel.edu/weeklycropupdate/?p=280 (accessed 28 Sept. 2014).

Tietz. 1976. The Farmer. February 7, 1976. p. 23.

Timonin, M.I. 1947. Microflora of the rhizosphere in relation to the manganese deficiency disease of oats. Soil Sci. Soc. Am. Proc. 11:284–292. doi:10.2136/sssaj1947.036159950011000C0054x

Tiwari, K.N., and B.S. Dwivedi. 1990. Response of eight winter crops to zinc fertilizer on a Typic Ustochrept soil. J. Agric. Sci. 115:383–387. doi:10.1017/S0021859600075821

Toker, C., T. Yildirim, H. Canci, N.E. Inci, and F.O. Ceylan. 2010. Inheritance of resistance to iron deficiency chlorosis in chickpea (*Cicer arietinum* L.). J. Plant Nutr. 33:1366–1373. doi:10.1080/01904167.2010.484096.

Touchton, J.T., and F.C. Boswell. 1975. Boron application for corn grown on selected southern soils. Agron. J. 67:197–200. doi:10.2134/agronj1975.00021962006700020006x

Tremblay, N., Y.M. Bouroubi, C. Belec, R.W. Mullen, N.R. Kitchen, W.E. Thomason, S. Ebelhar, D.B. Mengel, W.R. Raun, D.D. Francis, E.D. Vories, and I. Ortiz-Monasterio. 2012. Corn response to nitrogen is influenced by soil texture and weather. Agron. J. 104:1658–1671. doi:10.2134/agronj2012.0184

US Salinity Laboratory Staff. 1954. Diagnosis and improvement of saline and alkali soils. US Government printing office, Washington, DC.

van Berkum, P. and J.J. Fuhrmann. 2000. Evolutionary relationships among soybean bradyrhizobia reconstructed from 16S rRNA gene and internally transcribed spacer region sequence divergence. Int. J. Syst. Evol. Microbiol. 50:2165–2172. doi:10.1099/00207713-50-6-2165

Vanotti, M.B., and L.G. Bundy. 1995. Soybean effects on soil nitrogen availability in crop rotations. Agron. J. 87:676–680. doi:10.2134/agronj1995.00021962008700040012x

Vitosh, M.L., J.W. Johnson, and D.B. Mengel. 1995. Tri-state fertilizer recommendations for corn, soybeans, wheat and alfalfa. Michigan State Univ. Ext. Bull. E-2567. Michigan State Univ., East Lansing, MI.

Vyn, T.J., and K.J. Janovicek. 2001. Potassium placement and tillage system effects on corn response following long-term no-till. Agron. J. 93:487–495. doi:10.2134/agronj2001.933487x

Wakeel, A., D. Steffens, and S. Schubert. 2010. Potassium substitution by sodium in sugarbeet (*Beta vulgaris*) nutrition on K-fixing soils. J. Plant Nutr. Soil Sci. 173:127–134. doi:10.1002/jpln.200900270

Walley, F.L., S. Kyei-Boahen, G. Hnatowich, and C. Stevenson. 2005. Nitrogen and phosphorus fertility management for desi and kabuli chickpea. Can. J. Plant Sci. 85:73–79. doi:10.4141/P04-039

Ware, L.S. 1880. The sugarbeet including a history of the beet sugar industry in Europe. Henry Carey Baird and Co., Philadephia, PA.

Weir, R.G. 2004. Molybdenum deficiency in plants. AgFact AC4, New South Wales Department of Primary Industries. Orange, NSW.

Weisz, R., and R. Heiniger. 2004. Nitrogen management for small grains. In: R. Weisz, editor, Small grain production guide 2004–2005. Ext. Circ. AG-580. North Carolina Coop. Ext. Serv., Raleigh, NC. p. 21–24.

Wen, G., C. Chen, K. Neill, D. Wichman, and G. Jackson. 2008. Yield response of pea, lentil and chickpea to phosphorus addition in a clay loam soil of central Montana. Arch. Agron. Soil Sci. 54:69–82. doi:10.1080/03650340701614239

Westermann, D.T., D.W. James, T.A. Tindall, and R.L. Hurst. 1994. Nitrogen and potassium fertilization of potatoes: Sugars and starch. Am. Potato J. 71:433–453. doi:10.1007/BF02849098

Wetterauer, D.C., and R.J. Killorn. 1996. Fallow-and flooded-soil syndromes: Effects on crop production. J. Prod. Agric. 9:39–41. doi:10.2134/jpa1996.0039

Whitney, D.A. 2012. Micronutrients: Zinc, iron, manganese and copper. Chapter 9. In: Recommended chemical soil test procedures for the North Central Region. NCERA-13 publication. Missouri Agric. Exp. Station SB 1001, Columbia, MO. http://extension.missouri.edu/explorepdf/specialb/sb1001.pdf (revised 2012) (accessed 28 Sept. 2014).

Wilson, W.W., and D. Miljkovic. 2011. Dynamic inter-relationships in hard wheat basis markets. In: Proc. of the NCCC-134 Conf. on Applied Commodity Price Analysis, Forecasting, and Market Risk Management. St. Louis, MO. http://www.farmdoc.illinois.edu/nccc134.

Wortmann, C.S., A.R. Dobermann, R.B. Fertuson, G.W. Hergert, C.A. Shapiro, D.D. Tarkalson, and D.T. Walters. 2009. High-yielding corn response to applied phosphorus, potassium and sulfur in Nebraska. Agron. J. 101:546–555. doi:10.2134/agronj2008.0103x

Yousfi, S., M. Wissal, H. Mahmoudi, C. Abdelly, and M. Gharsalli. 2007. Effect of salt on physiological responses of barley to iron deficiency. Plant Physiol. Biochem. 45:309–314. doi:10.1016/j.plaphy.2007.03.013

Yusuf Ali, M., L. Krishnaumurthy, N.P. Saxena, O.P. Rupela, J. Kumar, and C. Johansen. 2002. Scope for genetic manipulation of mineral acquisition in chickpea. Plant Soil 245:123–124. doi:10.1023/A:1020616818106

Zohary, D., and M. Hopf. 1973. Domestication of pulses in the Old World. Science 182:887–894. doi:10.1126/science.182.4115.887

Zubriski, J.C. 1964. The effects of fertilizers on flax. H-15-8 Annual Report for 1963. Department of Soils, North Dakota State Univ., Fargo. p. 10–32.

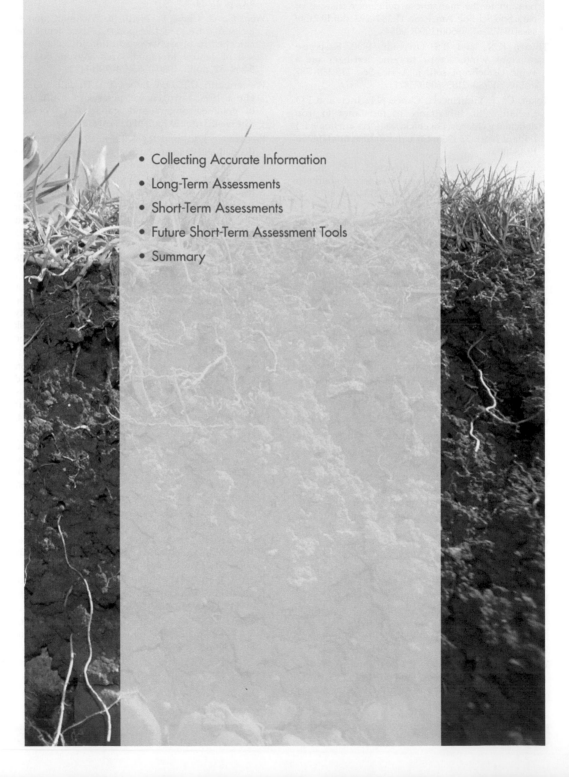

- Collecting Accurate Information
- Long-Term Assessments
- Short-Term Assessments
- Future Short-Term Assessment Tools
- Summary

Assessing a Fertilizer Program: Short- and Long-Term Approaches

David E. Clay,* Graig Reicks, Jiyul Chang, Tulsi Kharel, and Stephanie A.H. Bruggeman

Increasingly agronomists are asked to increase yields while simultaneously reducing fertilizer costs and the impact of agriculture on the environment. Achieving these goals is complicated by climatic variability that can reduce fertilizer efficiency and increase yield losses. One technique to improve fertilizer efficiency is to conduct long-term assessments of the fertilizer program. Specifically discussed here are the importance of field scouting, long-term N and P assessments, and future opportunities for identifying the yield limiting factors. Soil nutrient assessments provide information needed to implement locally adapted fertility programs.

Collecting Accurate Information

Agricultural fields are a mosaic of habitats, each having unique characteristics that influence soil properties and the ability of the crop plants to respond to the applied fertilizer. The effectiveness of a fertilizer application rests on the ability to develop appropriate solutions for each unique environment. All assessment techniques starts with collecting accurate information.

Soil Sampling

Soil sampling protocols are site, nutrient, and crop specific, and they can be collected following a wide variety of approaches, including grid cell, whole field, grid point, and management zone (Kitchen et al., 1990; Clay et al., 1997, 2002, 2011; Franzen and Cihacek, 1998; Stecker et al., 2001). Each approach has unique strengths and weaknesses and requires different skill levels. For example, collecting a single sample from a whole field has a low skill requirement, whereas identifying a management zone has a high skill requirement. For fertilizer

Abbreviations: ELISA, enzyme-linked immunosorbent assay; GPS, global positioning system; OM, organic matter; SOC, soil organic carbon.

David E. Clay, Graig Reicks (graig.reicks@sdstate.edu), Jiyul Chang (Jiyul.Chang@sdstate.edu), Tulsi Kharel (Tulsi.Kharel@sdstate.edu), and Stephanie A.H. Bruggeman (Stephanie.Bruggeman@sdstate.edu), Plant Science Dep., South Dakota State Univ., Berg Agricultural Hall, P.O. Box 2207A, Brookings, SD 57007. *Corresponding author (david.clay@sdstate.edu).

doi:10.2134/soilfertility.2015.0079

Soil Fertility Management in Agroecosystems
Amitava Chatterjee and David Clay, Editors

assessments, soil samples should be collected using appropriate protocols at a standard time. General observations for soil sampling might include the following:

1. Carefully consider when to collect the sample. Soil samples collected following harvest often have lower nutrient concentrations than those collected in the spring. Differences between the fall and spring samples might reveal something about the soils' ability to resupply nutrients.

2. Prior to sampling, identify the goals and where additional information is needed. Where possible, avoid collecting samples from guess rows, field entrances, field discontinuities (eroded and low areas), headlands and boarders, old homesteads, and animal confinement areas (Fig. 1).

3. Select a laboratory that provides an analysis that helps you achieve your goals (Barbagelata and Mallarino, 2013; Nathan and Gelderman, 2015). When selecting a laboratory, consider the reliability of the results as well as the turnaround time. Precision and accuracy represent two different terms. Precision is a measure of repeatability, whereas accuracy is a measure of the ability to obtain the correct value. Laboratories can be precise and inaccurate as well as imprecise and inaccurate. Where possible, select laboratories that are precise and accurate. The Soil Science Society of America sponsors the North American Proficiency Testing (NAPT) program that provides a certification of laboratories. A list of certified laboratories is available at http://www.naptprogram. org/ (accessed 1 Feb. 2016).

4. Use recommended techniques to prepare and ship the sample for analysis (Nathan and Gelderman, 2015).

Scouting Fields

When scouting the field, it is important to identify your sampling approach (Fig. 2) and determine the plant growth stage and population. When collecting tissue and soil samples, inspect the plants for nutrient deficiencies (Table 1). If tissue samples are collected, appropriate protocols should be followed. Data collected from scouting is useful in the development of management practices for future years. To facilitate this analysis, yield data, associated cultural practices, images, pest problems, personal notes, sampling dates and protocols, and soil test results should be placed into long-term storage. Choices for long-term storage include:

• Printed hard copies of all data from a given field.

• On-farm storage of the digital records. This is complicated by computer systems that routinely change.

• Off-farm storage by a data management company.

• Routinely update data to current data storage formats.

Pre-1984 Farm Site

Fig. 1. An aerial image and field soil test P contour map. Very high P levels can be found in old home sites.

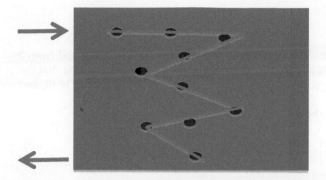

Fig. 2. During scouting walk in the field at least 10 steps from the field edge and examine 10 plants at every black dot.

Table 1. General plant deficiency symptoms that can be observed when scouting a field.

Nutrient	Symptom	Plant part	Solution
Nitrogen	general yellowing	older parts first	in legume treat seed with *Bradyrhizobium* or apply N fertilizer
Phosphorus	dark green or reddish purple leaves	older parts first	apply P fertilizer, check soil P level
Potassium	wilting, interveinal chlorosis, and scorching of leaf margins starting at the edge	older parts first	apply K fertilizer, check soil K levels
Sulfur	general yellowing	younger leaves first	apply S fertilizer and check soil S level
Iron	yellowing of veins of the leaves generally found in high pH soils; whole leaf may turn white	younger leaves first	use Fe efficient cultivars and treat seed with Fe
Zinc	pale green plants; interveinal mottling (or interveinal chlorosis in drybean) of older leaves leading to bronze necrosis; green veins	younger leaves first	apply Zn fertilizer

Long-Term Assessments

Nitrogen Assessment

Field crops obtain N from fertilizer, N fixation, and the mineralization of organic matter. In a simple sense, the N contained within the fertilizer is taken up by the plant, immobilized, or lost through leaching or denitrification (Clay et al., 1990b), whereas the N contained in the organic matter is make available to the plant through biological activity (Shaver, 2014). Because the amount of soil organic matter is directly related to productivity and resilience (Clay et al., 2014), it has been used as a proxy for estimating the impact of the agricultural system on soil health (Clay et al., 1990a,b).

Soil organic matter and N long-term assessments provide valuable information about soil health and resilience. Increases in soil organic matter indicates the soils long-term productivity

potential has increased and its ability to help plant overcome adverse climate conditions has been improved (Clay et al., 2014). In addition, soil organic matter contains N as well as other nutrients that can be made available to growing plants. If the assessment indicates that soil organic matter has decreased, this indicates that soils yield potential and resilience has diminished. Soil organic matter can be rebuilt by reducing the tillage intensity, growing cover crops, and returning the crop residues to the field (Clay et al., 2015). Sample calculations for conducting an N assessment across the agricultural system are provided in Problem 1 (Box 1). This assessment shows that increasing the soil organic matter from 2 to 2.5 % resulted in a 544 kg N ha^{-1} (488 lbs N a^{-1}) increase the amount of organic N. The N contained in the SOC could be derived from many sources, including immobilization of applied fertilizer, N fixation

Box 1.

Problem 1. If your soil organic matter (OM) in the surface 6 in (15 cm) has increased from 2 to 2.5%, how much additional C and N is stored in the soil? In this calculation, assume that the surface 6 in contains 1.875 million kg soil ha^{-1} (1.68 million lb soil a^{-1}), the C/N ratio is 10, and OM is 58% carbon.

Step 1. Determine the amount of C in the soil

$$\left(\frac{0.025-0.02 \text{ kg OM}}{\text{kg soil}}\right) \times \frac{1,875,000 \text{ kg soil}}{\text{ha}} = 9,375 \text{ kg OM/ha}$$

In this calculation kg ha^{-1} can be converted to lb a^{-1} by using 1.68 million lb soil a^{-1} and changing kg to lb

$$\frac{(0.025-0.02 \text{ lb OM})}{\text{lb soil}} \times \frac{1,680,000 \text{ lb soil}}{\text{acre}} = 8,400 \text{ lb/acre}$$

Step 2. Convert to organic C

$$\frac{9375 \text{ kg OM}}{\text{ha}} \times \frac{0.58 \text{ kg C}}{1 \text{ kg OM}} = 5440 \text{ kg C/ha}$$

$$\frac{8400 \text{ lb}}{\text{acre}} \times \frac{0.58 \text{ lb C}}{1 \text{ lb organic matter}} = 4870 \text{ lb/a}$$

Step 3. Calculate change in soil N

$$\frac{5440 \text{ kg C}}{\text{ha}} \times \frac{0.1 \text{ kg N}}{1 \text{ kg C}} = 544 \text{ kg N/ha}$$

$$\frac{4870 \text{ lb}}{\text{acre}} \times \frac{0.1 \text{ lb N}}{1 \text{ lb C}} = 487 \text{ lb N/acre}$$

by legume plants, atmospheric N deposition, or the use of cover crops.

Conducting a Long-Term Phosphorus Assessment

Phosphorus assessments are based on determining changes in soil test P, plant P uptake, and P additions. The effectiveness of the P fertilizer program can be assessed by tracking changes in soil P over time because it's relatively immobile in soil. When making this comparison, spring soil P levels are often higher than fall values. To determine temporal changes in soil P, the soil samples must be collected by common protocols from similar locations and dates. This may require the use of a global positioning system (GPS) to identify the sample location. Changes in the soil test values

can provide critical information about the soil nutrient status. When considering soil nutrient changes, it is important to consider that factors other than plant removal may impact the soil test values. For example, soil test P can increase under anaerobic conditions and decrease under aerobic conditions. This change is attributed to changes in the oxygen concentration and associated changes in the relative amounts of Fe^{+2} (anaerobic) and Fe^{+3} (aerobic) contained in the soil solution. In a second example, soil test K can be reduced when K^+ is immobilized into by soil clays such as vermiculite and mica. Changes is the soil test value may also result from the fertilizer additions exceeding the removal amounts. Regardless of the factor responsible for the change, tracking soil nutrients over time provide critical information for assessing the fertilizer effectiveness.

Assessment information can help the producer apply the right fertilizer at the right time, at the right location, at the right rate (Fixen, 2007). The assessment is a two-step process where nutrient removal is compared with nutrient additions (budget). The second step is to compare the budget with changes in the soil test value. Soil nutrient information can be used by the land manager to develop their locally based fertilizer program. For example, the implementation of the soil P assessment would be different for soils containing between 0 and 20 mg P g^{-1} soil than those containing >50 mg P g^{-1} soil.

Determining Phosphorus Removal

Yield monitor data, when combined with average nutrient levels in grain and tissue samples, can be used to assess soil nutrient mining and help determine if a change in the fertilizer program is warranted (Fig. 3). If the soil nutrient concentrations are much higher than the critical level, consider reducing the rate or not applying any fertilizer. In some situations, environmental considerations necessitate decreasing or eliminating additional P applications.

When making nutrient budget calculations, all crops used in the rotation must be considered. Removal rates for selected crops are provided in Table 2 and Clay et al. (2011). Removal can be converted from P to P$_2$O$_5$ by dividing the removal value by 0.436, and K can

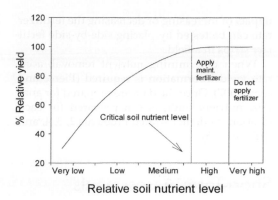

Fig. 3. A conceptual relationship between relative yield and relative soil nutrient level. This chart shows the relationship between the critical soil nutrient level, maintenance fertilizer applications, and where not to apply any additional fertilizer.

be converted to K$_2$O by dividing the removal value by 0.83.

The amount of nutrient removed from a field is determined by summing the amount of nutrients removed by the harvested crop over several years, whereas additions are determined by summing the nutrient additions, including manure (Problem 2 and 3). If the analysis indicates that mining occurred (outputs > inputs) and soil test values have decreased below the critical nutrient level, consider increasing the fertilizer rate. The potential

Table 2. Estimates of nutrient removal of N, P, K, Mg, and S by major South Dakota crops (Clay et al., 2011). The units are pounds per bushel and kilograms per megagram.

Crop	Plant part	Unit	N	P	K	Mg	S
Corn	Grain	lb/bu	0.9	0.17	0.22	0.09	0.08
	Grain	kg/Mg	16.1	3.0	3.93	1.6	1.4
	Stover	lb/ton	16	2.5	33.2	5	3
	Stover	kg/Mg	8	1.3	16.6	2.5	1.5
Soybean	Grain	lb/bu	3.8	0.37	1.1	0.21	0.18
	Grain	kg/Mg	63.3	6.2	18.0	3.5	3.0
	Stover	lb/ton	40	3.9	31	8.1	6.2
	Stover	kg/Mg	20	1.95	15.5	4.1	3.1
Wheat	Grain	lb/bu	1.5	0.26	0.28	0.15	0.1
	Grain	kg/Mg	25.0	4.4	4.7	2.5	1.7
	Straw	lb/ton	14	1.44	19.9	2	2.8
	Straw	kg/Mg	7.0	0.7	10.0	1.0	1.4

impact of increasing or decreasing the fertilizer rate can be tested by placing side-by-side fertilizer strips in a field.

When determining nutrient removal, accurate yield information is required (Pierce and Clay, 2007). Once the data are prepared for analysis, nutrient budgets can prepared for each location, as demonstrated Problems 2, 3, 4, and 5 (Boxes 2–5).

Short-Term Assessments

A problem with correctly identifying the yield-limiting factor is that a plant's response to one stress may influence its ability to respond to other stresses (Bloom et al., 1985; Kim et al., 2008; Kharel et al., 2011; Hansen et al., 2013). Hansen et al. (2013) used ^{13}C isotopic discrimination, plant tissue samples, soil nutrient analysis, and gene expression (transcriptome analysis) to investigate this question. This work showed that in response to water stress, corn downregulated several genes associated with nutrient uptake, photosynthesis, and plant defenses against pests. Changes in gene expression may have influenced the nutrient uptake capacity and the ability of the plant to respond to applied fertilizer. This work suggests that tissue P concentrations were influenced by both soil P levels and water stress. Reese et al. (2014) had similar results and reported that water used by winter cover crops can negatively impact genes associated with nutrient uptake. Similar findings have been observed with N (Rubio et al., 2003; Kim et al., 2008; Kharel et al., 2011).

Stalk Nitrate Test: In-Season Assessment

Previous research has shown that the stalk nitrate concentrations are influenced by N rate and that high values can result from the N rate exceeding the plant requirement. However, the interpretation of the stalk nitrate test may be hybrid and location specific. In this test, 8-in stalk segments (6 to 14 in segment) are collected from the field 2 or 3 wk before black layer (Camberato and Nielsen, 2014). In Indiana, if the nitrate-N concentration is <450 mg g^{-1} (ppm), the plant is likely N limited, whereas if the concentration is >2000 mg g^{-1}, the plant is not N limited (Camberato and Nielsen, 2014). Slightly different values are suggested for Minnesota where the adequate levels are defined as between 700 and 2000 ppm (Kaiser et al., 2013).

Tissue Sampling

Tissue sampling techniques and interpretation are crop growth age, plant, and location specific, and when combined with associated information they can be used to help identify the yield-limiting factors (Flynn et al., 1999; Mills and Jones, 1996). When considering tissue sampling, check with the local laboratory about sampling protocols. In tissue sampling, samples collected from a prescribed location are analyzed for a range of nutrients. The protocols were based on research and determined the relationship between plant part, growth stage, nutrient concentration, and yield. In Minnesota, for soybeans it is recommended that entire plants are collected at the seedling growth stage, whereas for plants between the R1 to

---

Box 2.

Problem 2. Estimating crop P removal in a corn and soybean rotation.

Phosphorus removed by 60 bu soybeans/acre and 200 bu corn/acre:

$$\text{Pounds of P/acre removed by a 60 bu/a soybean crop} = \frac{60 \text{ bu}}{\text{acre}} \times \frac{0.37 \text{ lb P}}{\text{bu}} = \frac{22.2 \text{ lb P}}{\text{acre}}$$

$$\text{Pounds of P/acre removed by a 200 bu/acre corn crop} = \frac{200 \text{ bu}}{\text{acre}} \times \frac{0.17 \text{ lb P}}{\text{bu}} = \frac{34 \text{ lb P}}{\text{acre}}$$

Total removal in a corn and soybean rotation is 56 lb P/acre or 128 lb P_2O_5/acre (to convert from P to P_2O_5, multiply P by 2.29 or divide P by 0.436). Pounds/acre are converted to kg/ha by converting the units.

$$\frac{34 \text{ lb P}}{\text{acre}} \times \frac{2.47 \text{ acre}}{\text{ha}} \times \frac{\text{kg}}{2.205 \text{ lb}} = \frac{38 \text{ kg}}{\text{ha}}$$

Box 3.

Problem 3. Determine the amount of N and P removed in the grain of a 10 Mg ha⁻¹ (10 metric tons ha⁻¹) corn crop and a 2 Mg ha⁻¹ soybean crop. Note in this calculation 1000 kg = 1 Mg ha⁻¹.

N removed by corn and soybean

$$\text{Corn N} = \frac{10 \text{ Mg}}{\text{ha}} \times \frac{16.1 \text{ kg N}}{\text{Mg}} = \frac{160 \text{ kg N}}{\text{ha}}$$

$$\text{Soybean N} = \frac{2 \text{ Mg}}{\text{ha}} \times \frac{620 \text{ kg}}{\text{Mg}} = \frac{40 \text{ kg N}}{\text{acre}}$$

$$\text{Total N} = 160 + 40 = 200 \text{ kg N/ha}$$

P removed by corn and soybean

$$\text{Corn P} = \frac{10 \text{ Mg}}{\text{ha}} \times \frac{1.3 \text{ kg P}}{1 \text{ Mg}} = \frac{13 \text{ kg P}}{\text{ha}}$$

$$\text{Soybean P} = \frac{2 \text{ Mg}}{\text{ha}} \times \frac{1.95 \text{ kg P}}{\text{Mg}} = \frac{3.9 \text{ kg P}}{\text{ha}}$$

$$\text{Total P}_2\text{O}_5 = 13 + 3.9 = 16.9 \text{ kg P/ha}$$

The N and P budget were determined by subtracting removal rates from additions. The N budgets have little meaning because soybeans have the capacity to convert atmospheric N_2 to plant N (N fixation). For P, if the budget is negative (removal > additions), check the soil test and plant nutrient values. If the soil test values have decreased and the plant P level is below the critical P level, consider increasing the amount of P applied.

Box 4.

Problem 4. Determine the amount of N and P_2O_5 harvested in a 200 bu/acre corn crop. A 200 bu corn crop produces approximately 9464 lb of dry stover [(200 bu/acre) × (47.32 lb dry grain/bu grain) × (1 lb stover/lb dry grain)]. This calculation assumes a harvest index [HI = dry grain/(grain + stover)] = 0.50, and that to convert P to P_2O_5 divide P by 0.43611.

N and P_2O_5 in the Grain + Stover

$$\text{Corn N} = \frac{200 \text{ bu}}{\text{acre}} \times \frac{0.9 \text{ lb N}}{\text{bu}} + \frac{9464 \text{ lb stover}}{\text{acre}} \times \frac{\text{ton}}{2000 \text{ lb}} \times \frac{16 \text{ lb}}{1 \text{ ton}} = \frac{256 \text{ lb N}}{\text{acre}}$$

$$\text{Corn P}_2\text{O}_5 = \frac{200 \text{ bu}}{\text{acre}} \times \frac{0.38 \text{ lb P}_2\text{O}_5}{\text{bu}} + \frac{9464 \text{ lb stover}}{\text{acre}} \times \frac{\text{ton}}{2000 \text{ lb}} \times \frac{5.73 \text{ lb}}{1 \text{ ton}} = \frac{103 \text{ lb}}{\text{acre}}$$

Box 5.

Problem 5. The soil test P value in 2009 is 20 ppm, and the soil test value in 2015 is 18 ppm. During this 6-yr time period, you have applied 80 lb P_2O_5 a⁻¹ annually and harvested a 200 bu a⁻¹ corn crop annually. What do you do?

The soil test value indicate that soil test P has gradually decreased and based on Problem 4, your nutrient removal values exceed the P additions. On the basis of this analysis the producer should consider increasing the P rate to 103 lb P_2O_5 a⁻¹.

If the soil test P value decreased from 60 to 50 ppm, how would you change your recommendation?

R3 growth stages it is recommended that 30 to 50 of the most recently mature trifoliates are collected (Table 3; Kaiser et al., 2013).

For corn grown in Minnesota, 15 to 20 whole plants should be collected at the seedling stage, whereas for plants between 12-in tall (30 cm) to tasseling, collect 15 to 20 of the first fully developed leaves from the top of the plant. For plants between tasseling and silking, collect 12 to 20 of the leaves directly below the ear (Kaiser et al., 2013). Again, care should be followed to make sure the plants do not mold. The expected nutrient ranges for corn are provided in Table 4.

In Kansas, tissue sampling for hard red wheat consists of collecting whole plants (above ground only) from tillering to jointing and collecting the flag leaf from booting to heading (Mengel, 2015). Each sample should consist of subsamples containing at least 20 plants or leaves. The sufficiency ranges for the major nutrients are provided in Table 5.

Models

Increasingly, crop models are being used to integrate soil, plant, and climatic conditions into fertilizer recommendation (Setiyono et al., 2011). For example, Setiyono et al. (2011) developed and tested a robust N fertilizer decision-support tool. The research showed that the model was sensitive to changes in soil properties, climatic conditions, crop rotations, tillage, fertilizer formulations, and fertilizer

Table 3. Expected ranges for soybean trifoliates in Minnesota collected between R1 and R3 (modified from Kaiser et al., 2013).

Plant nutrient	Unit	Expected range
Nitrogen	%	4.26–5.5
Phosphorus	%	0.26–0.50
Potassium	%	1.71–2.50
Calcium	%	0.36–2.00
Magnesium	%	0.26–1.00
Iron	ppm	51–350
Zinc	ppm	20–50
Boron	ppm	21–55
Copper	ppm	10–30
Manganese	ppm	21–100

Table 4. Expected ranges of selected nutrients for corn collected at three growth stages (modified from Kaiser et al., 2013).

Nutrient	Unit	Seedling	Vegetative	Tasseling to silking
Nitrogen	%	4.0–5.0	3.5–4.5	2.76–3.75
Phosphorus	%	0.4–0.6	0.35–0.50	0.25–0.50
Potassium	%	3.0–5.0	2.0–3.5	1.75–2.75
Calcium	%	0.51–1.6	0.20–0.80	0.30–0.60
Magnesium	%	0.3–0.6	0.20–0.60	0.16–0.40
Sulfur	%	0.18–0.40	0.18–0.40	0.16–0.40
Iron	ppm	40–500	25–250	50–250
Zinc	ppm	25–60	20–60	17–75
Boron	ppm	6–25	6–25	5.1–40
Manganese	ppm	40–160	20–150	50–250
Copper	ppm	6–20	6–20	3–15

Table 5. Expected sufficiency ranges of wheat in Kansas (modified from Mengel, 2015).

Nutrient	Unit	Whole plant, tillering to jointing	Flag leaf, boot to heading
Nitrogen	%	3.5–4.5	3.5–4.5
Phosphorus	%	0.3–0.5	0.3–0.5
Potassium	%	2.5–4.0	2.0–3.0
Calcium	%	0.2–0.5	0.3–0.5
Magnesium	%	0.15–0.5	0.2–0.6
Sulfur	%	0.19–0.55	0.5–0.55

application timing. Increasingly, models are being used as decision support tools.

Future Short-Term Assessment Tools

Molecular Techniques

Molecular techniques are being explored for their ability to provide a rapid early season assessment of nutrient shortages. However, this area of work is in its infancy. Currently microarrays, real-time quantitative PCR, and next-generation RNA sequencing are used to measure the influence of biotic and abiotic stress on plant productivity (Hansen et al., 2013; Reese et al., 2014). However, these tools are expensive and cannot be used for real-time decisions.

Research is being conducted to convert these techniques into a rapid assessment tool. Developing this basic knowledge is confounded by laboratory molecular techniques that are rapidly changing (Table 6).

Remote Sensing

Remote sensing information can be collected by unstaffed aerial vehicles (drones), planes, and spaced-based satellites. Each platform has different advantages and disadvantages (Table 7). Understanding the benefits and limitations of various platforms and the sensors is critical for selecting the appropriate sensor. In a general sense, resolution is directly related to cost.

Remote sensing data consists of different bands, each from a different wavelength, and pixels with a specified spatial resolution. The

Table 6. Pros and cons of current molecular techniques.

Molecular technique	Pros	Cons	Summary
Microarray	less expensive than RNA Sequencing, generates less data on specific genes	very technical, experimental design can make or break results, generates a lot of data	method was a stepping stone into RNA Sequencing, and is now only used in specific instances
RNA sequencing	very detailed, can gives a complete picture of all genes involved	expensive, experimental design can make or break experiment, generates an immense amount of data to be analyzed	great method to answer key questions about how plants work in specific/controlled field situations
RT-PCR	faster than microarrays or RNA Sequencing, accurate	highly technical, need specially equipped laboratory	best for a relatively quick screen of select genes for expression
ELISA	easy to use, like a pregnancy test, quick results	complicated to develop, very few tests based on plant related proteins/gene expression	will be great in the future, but not yet

Table 7. Advantages and disadvantages of hand or machine mounted, aerial, and space-based platforms for remote sensing data collection.

Platform	Advantages	Disadvantages
Hand or machine	identifies the reflectance characteristics, flexible availability, can be used for real-time spraying, application, can be mounted on an applicator	data collected from a single point
Drone	can provide high resolution, can select different sensors, may have a low cost	unstable platform that can produce distortions and slow data processing. May require certification
Aircraft	may be relatively flexible and provide high resolution	High cost and data processing may slow data processing
Satellite	cost can range from free to very expensive, limited by cloud cover, stable platform, and many archived data sets	clouds may hide ground features, has a fixed schedule, cloud cover can limit usefulness, and data collection and plant requirements may not match

primary advantages of remote sensing are that large amounts of information can be quickly processed into a single image, whereas the primary disadvantages are cost, multiple stresses may have similar impacts on reflectance, and turnaround time (Clay et al., 2006, 2012; Mishra et al., 2008). Many agricultural applications of remote sensing are discussed in Reese (2004).

Remote sensing data can be visualized and processed in a variety of ways. For example, reflectance from the primary blue, green, and red bands can be processed into a true color composite image, while reflectance in green, red, and near-infrared bands can be used to produce a false color composite image. In addition, the bands can also be combined to produce indices. One of the most commonly used indices is the normalized difference vegetation index (NDVI), which is calculated using the equation NDVI = (near-infrared − red)/(near-infrared + red). The NDVI values can range from −1 to 1. The NDVI is widely used in agriculture, and research is being conducted to assess the usefulness of remote sensing for controlling in-season N applications.

Summary

Increasingly, agronomists are asked to consider production and environmental goals simultaneously (Fixen, 2007). Soil and plant assessments are critical components of achieving this goal. In the future these time tested assessment techniques may be augmented by complex mathematical models and molecular based assessment tools that accurately identify the yield limiting factors.

References

Barbagelata, P.A., and A.P. Mallarino. 2013. Field correlations of potassium soil test methods based on dried and filed-moist soil samples for corn and soybeans. Soil Sci. Soc. Am. J. 77:318–327. doi:10.2136/sssaj2012.0253

Bloom, A.J., F.S. Chapin, III, and H.A. Mooney. 1985. Resource limitation in plants-an economic analogy. Annu. Rev. Ecol. Syst. 16:363–392. doi:10.1146/annurev.es.16.110185.002051

Camberato, J., and B. Nielsen. 2014. Corn stalk nitrate test-research and recommendation update. Purdue Univ. https://www.agry.purdue.edu/ext/soilfertility/news/cornstalknitrate.pdf (accessed 26 Oct. 2015).

Clay, D. E., C.G. Carlson, K. Brix-Davis, J. Oolman, and R. Berg. 1997. Soil sampling strategies for estimating residual nitrogen. J. Prod. Agric. 10:446–452. doi:10.2134/jpa1997.0446

Clay, D.E., C.E. Clapp, J.A. Molina, and R.H. Dowdy. 1990a. Influence of nitrogen fertilization, tillage, and residue management on an N mineralization index. Commun. Soil Sci. Plant Anal. 21:323–335.

Clay, D.E., S.A. Clay, C.G. Carlson, and S. Murrell. 2011. Mathematics and Calculations for Agronomists and Soil Scientists. International Plant Nutrition Institute. http://www.ipni.net/article/IPNI-3227 (accessed 26 Oct. 2015).

Clay, D., S. Clay, K. Reitsma, B. Dunn, A. Smart, G. Carlson, D. Horvath, and J. Stone. 2014. Does the conversion of grasslands to row crop production in semi-arid areas threaten global food security? Global Food Security 3:22–30.

Clay, D.E., T.P. Kharel, C. Reese, D. Beck, C.G. Carlson, S.A. Clay, and G. Reicks. 2012. Winter wheat crop reflectance and N sufficiency index values are influenced by N and water stress. Agron. J. 104:1612–1617. doi:10.2134/agronj2012.0216

Clay, D.E., K.-I. Kim, J. Chang, S.A. Clay, and K. Dalsted. 2006. Characterizing water and N stress in corn using remote sensing. Agron. J. 98:579–587. doi:10.2134/agronj2005.0204

Clay, D.E., N. Kitchen, C.G. Carlson, J.L. Kleinjan, and W.A. Tjentland. 2002. Collecting representative soil samples for N and P fertilizer recommendations. Crop Manage. 1:1. doi:10.1094/CM-2002-1216-01-MA.

Clay, D.E., G.L. Malzer, and J.L. Anderson. 1990b. Tillage and dicyandiamide influence on nitrogen fertilizer immobilization, remineralization, and utilization by maize. Biol Fertil. Soil 9: 220–225.

Clay, D.E., G. Reicks, C.G. Carlson, J. Miller, J.J. Stone, and S.A. Clay. 2015. Tillage and corn residue harvesting impacts surface and subsurface carbon sequestration. J. Enron. Qual. 44:803–809 doi:102134/jeq2014.070322

Fixen, P. 2007. Can we define a global framework within which fertilizer best management practices can be adapted to local conditions? In: Fertilizer best management practices. Intl. Fertilizer Industry Assoc., Paris.

Flynn, R., S.T. Ball, and R.D. Baker. 1999. Sampling for plant tissue analysis. New Mexico State Univ.. Guide A-123. http://aces.nmsu.edu/pubs/_a/A123/ (accessed 26 Oct. 2015).

Franzen, D., and L.J. Cihacek. 1998. Soil sampling as a basis for fertilizer application. SF-990. North Dakota State Univ. Ext. Serv. http://www.sbreb.org/brochures/soilsampling/soilsamp.htm (accessed 26 Oct. 2015).

Hansen, S., S.A. Clay, D.E. Clay, C.G. Carlson, G. Reicks, Y. Jarachi, and D. Horvath. 2013. Landscape features impacts on soil available water, corn biomass, and gene expression during the late vegetative growth stage. Plant Gen. 6:1–9. doi:10.3835/plantgenome2012.11.0029

Kaiser, D.E., J.A. Lamb, and C. Rosen. 2013. Plant analysis and interpretation. FO-7176-B (revised). Univ. of Minnesota Ext., St. Paul, MN. http://www.extension.umn.edu/agriculture/nutrient-management/docs/AG-FS-3176-1.pdf. (accessed 26 Oct. 2015).

Kharel, T.P., D.E. Clay, S.A. Clay, D. Beck, C. Reese, G. Carlson, and H. Park. 2011. Nitrogen and water stress affect winter wheat yield and dough quality. Agron. J. 103:1389–1396. doi:10.2134/agronj2011.0011

Kim, K.-I., D.E. Clay, C G. Carlson, S.A. Clay, and T. Trooien. 2008. Do synergistic relationships between nitrogen and water influence the ability of corn to use nitrogen derived from fertilizer and soil? Agron. J. 100:551–556. doi:10.2134/agronj2007.0064

Kitchen, N.R., J.L. Havlin, and D.G. Westfall. 1990. Soil sampling under no-till banded phosphorus. Soil Sci. Soc. Am. J. 54:1661–1665. doi:10.2136/sssaj1990.03615995005400060026x

Mengel, D. 2015. Recommendation for plant analysis of wheat. K-State Updates 506, April. https://webapp.agron.ksu.edu/agr_social/eu_article.throck?article_id = 535 (accessed 27 Oct. 2015).

Mills, H.A., and J.B. Jones, Jr. 1996. Plant analysis handbook 2: A practical sampling, preparation, analysis, and interpretation guide. MicroMacro Publ., Athens, GA.

Mishra, U., D. Clay, T. Trooien, K. Dalsted, and D.D. Malo. 2008. Using remote sensing based ET maps to assess landscape processes impacting soil properties. J. Plant Nutr. 31:1188–1202. doi:10.1080/01904160802134491

Nathan, M., and R. Gelderman. 2015. Recommended chemical soil test procedures for the north central region. #221 (revised). Missouri Agric. Exp. Stn. SD 2001. http://extension.missouri.edu/explorepdf/specialb/sb1001.pdf (accessed 26 Oct. 2015).

Pierce, F.J., and D. Clay. 2007. GIS applications in agriculture. CRC Press, New York.

Reese, C. 2004. Using remote sensing to make insurance claim for crop damage due to hail event. UMAC Success Story. http://www.umac.org/agriculture/ss/ganinsuranceclaimforcropdamageduetohailevent/detail.html. (accessed 27 Oct. 2015).

Reese, C., D.E. Clay, S.A. Clay, A. Bich, A. Kennedy, S. Hansen, and J. Miller. 2014. Wintercover crops impact on corn production in semi-arid regions. Agron. J. 106:1479–1488. doi:10.2134/agronj13.0540

Rubio, G., J. Zhu, and J.P. Lynch. 2003. A critical test of the two prevailing theories of plant responses to nutrient availability. Am. J. Bot. 90:143–152. doi:10.3732/ajb.90.1.143

Setiyono, T.D., H. Yang, D.T. Walters, A. Dobermann, R.B. Ferguson, D.F. Roberts, D.J. Lyon, D.E. Clay, and K.G. Cassman. 2011. Maize-N: A decision tool for nitrogen management in maize. Agron. J. 103:1276–1283. doi:10.2134/agronj2011.0053

Shaver, T., 2014. Nutrient management for agronomic crops in Nebraska. Univ. of Nebraska EC 155. available at http://extensionpublications.unl.edu/assets/pdf/ec155.pdf (accessed 21 Dec. 2015).

Stecker, J.A., J.R. Brown, and N.R. Kitchen. 2001. Residual phosphorus distribution and sorption in starter fertilizer band applied in no-till culture. Soil Sci. Soc. Am. J. 65:1173–1183. doi:10.2136/sssaj2001.6541173x

Introduction to Conceptual Models, Calculating and Using Rate Constants, Economics, and Problem Solving

Gregg Carlson, David E. Clay,* and Sharon A. Clay

Abstract

Mathematical models are used to extend research findings beyond the boundary conditions of the original experiment. Most modeling approaches are based on visualizing and developing conceptual diagrams, conducting focused experiments, creating the mathematics, and testing the model. Each modeling component leads to a better understanding of the problem. This chapter discusses each of these components as well as provides (i) a discussion on how different conceptual models can result in different experimental treatments and possible interpretations, (ii) examples on how to calculate and use rate constants to solve agricultural problems, and (iii) examples on how to calculate the economic optimum fertilizer rate.

Conceptual Models and Experimental Treatments

In soil fertility, models are used for a variety of purposes ranging from calculating fertilizer rates to estimating the N release from slow-release fertilizers (Setiyono et al., 2011). In all models, there are tradeoffs between simplicity and assumptions. For example, one of the most widely used models is,

N rate = k(yield goal) − N credits

The strengths and weaknesses of this model were associated with its simplicity (O'Neill et al., 2004; Kim et al., 2008, 2013). Simplicity makes the model easy to understand while simultaneously making it insensitive to environmental change. Adding additional terms to the model increases its sensitivity and complexity. Unfortunately, complexity does not always improve accuracy (Tebaldi

Abbreviations: SOC, soil organic carbon.

Gregg Carlson (gregg.carlson@sdstate.edu), David E. Clay, Sharon A. Clay (sharon.clay@sdstate.edu), Plant Science Dep., South Dakota State Univ., Brookings, SD 57007. *Corresponding author (david.clay@sdstate.edu).

doi:10.2134/soilfertility.2016.0002

et al., 2006; Camargo and Kemanian, 2016). For example, Cramer et al. (1999) reported that different climate models produced net primary productivity values that ranged from −66.3 to 44.4 Pg C yr[-1].

One of the first steps in model building is to build a conceptual model based on your observations. For example, you observe that a broken wooden fence had preferentially decomposed near the soil surface. Based on this visualization, a conceptual diagram is developed that indicates that the broken fence pose was the result of decreasing O_2 concentrations and microbial activity with increasing soil depth (Clay et al., 2012). Based on your model, oxygen sensors are installed at different soil depths.

Different visualizations can result in different hypothesizes, interpretations, and mathematical expressions. For example, Lehmann and Kleber (2015) outline two contrasting views of C turnover. One view of modeling carbon turnover was based on separating soil organic C into multiple pools (Adair et al., 2008), whereas the contrasting view was that because soil organic carbon (SOC) contains a continuum of pools, it should be modeled as a single pool. Based on these fundamental differences, the Century model separated SOC into multiple pools, while C-Farm used a single C pool (Parton et al., 1993; Kemanian and Stöckle, 2010).

The visualization of the problem may also affect the experimental design and ultimately the interpretation. For example, as the plant grows, it takes up nutrients and fixes C. The amount and where these nutrients are taken up is function of the plant's genetics and population. If two plants are growing next to each other, it has been visualized that these plants are competing for water, nutrients, and light (Clay et al., 2009; Moriles et al., 2012), and that the better competitor grows larger. Mathematically, this could be defined as yield = f(water, nutrient, light). An experiment designed to measure nutrient uptake under different populations would define plant growth as a function of changes in amount of water, nutrients, and light.

This view has been challenged by the observation that factors other than water, nutrients, and light may limit growth during the plant's weed-free period (Clay et al., 2009; Moriles et al., 2012). One factor that has been explored is that plants have the ability to sense the proximity of their neighbors through changes in the red/near-infrared ratio of reflected light. Mathematically, this could be defined by the function, yield = f(water, nutrient, light, genetic characteristics,

population). These two hypothesizes are fundamentally different and would produce different experiments and interpretations. One experiment would quantify the impact of population on yield over multiple plants, whereas the second experiment, would quantify interactions between two adjacent plants.

Once the problem is visualized and the research is completed, the relational diagram and associated computer code can be developed using software packages such as MATLAB's Simulink (MathWorks, 2015). A relational diagram shows the linkages and possible mathematical functions linking the various components of the model (Hansen et al., 1991; Halpin and Morgan, 2008).

Building the Mathematics

A number of different approaches can be used to develop the mathematics that defines the relationship between multiple factors. Two widely used approaches are mechanistic- and statistical-based models. Many mechanistic-based models are based on zero- and first-order kinetics, whereas statistical-based models may statistically define the empirical relationship between the various factors. Examples for both types of models are provided below.

Zero-Order Kinetics

Zero-order kinetics can be used to define many problems. In zero-order kinetics, the rate is independent of the substrate concentration. Zero-order kinetics have been used to define a number of processes including soil organic N mineralization (Boyle and Paul, 1989; Nira and Nishimune, 1993). In zero-order kinetics, the half-life is the length of time for one-half of the substrate (S_0) added to the soil at time 0 to be consumed. The half-life is calculated with the following equation:

$$\text{Half-life} = S_0/(2m)$$

where, S_0 is the amount of a substrate at time 0 in the soil, and m is slope of the linear equation relating the amount of substrate in the soil vs. time (See Box 1).

First-Order Kinetics

In first-order kinetics, the rate of change is dependent on the amount of substrate that is in the soil. First-order kinetics can be used to model many biological processes. For example, first-order kinetics are used to assess the suitability of seeding a cover crop into soil treated with a pre-emergence herbicide, determine N release from a fertilizer granule, and assess soil health (Clay et al., 2010). A characteristic of first-order kinetics is that the rate of change decreases with a decreasing substrate concentration. First-order kinetics is based on the following equation:

$$\frac{dy}{dt} = -ky$$

This equation means that the rate of change in y (dy) with respect to time (dt) is dependent on the rate constant k and the amount of y in the system. When integrated, the equation becomes

$$y_t = y_0 e^{-kt}$$

and by taking the natural logarithm of both sides the equation,

$$\ln[y_t] = \ln\left[y_0 e^{-kt}\right]$$

is derived. This equation can be arranged into the equation,

$$\ln(y_t) = -kt + \ln(y_0)$$

The residence time and half-life are calculated with the equations,

$$t_{residence} = \frac{1}{k}$$

$$t_{half\text{-}life} = \frac{\ln 2}{k}$$

This decreasing function, $y_t = y_0 e^{-kt}$, can be converted to a function that slowly increases to a maximum value using the equation, $y_t = y_{max}$ $(1 - e^{-kt})$. In this equation, y_{max} is maximum value (Box 2).

Validation

The confidence of extending the model beyond the boundary conditions of the experiment is derived by testing the model using a data set that was not used to build the model. Model validation can include multiple steps including testing the assumption and comparing the model predictions with those obtained from an independent data set (Box 3). One of the most common statistical techniques is to compare the predicted and measured values. In this analysis, a linear model is often used to compare the predicted and measured values. A high R^2 value and an intercept of 0 and slope of 1 does not guarantee that the model is not biased. In addition to conducting the regression analysis, the residuals should be inspected.

Box 1.

Example: In an experiment where nitrate concentration with time is monitored, the following data was collected. What is the zero-order rate constant?

Days	Nitrate-N
1	100
3	80
5	75
7	50
10	40
15	20

To solve this equation, nitrate-N was plotted against time, and a linear equation was determined. In this problem, $y = 99.5 - 5.65(time)$, $r^2 = 0.95$. The rate constant is 5.65 g day^{-1}. Multiple factors could be responsible for this hypothetical example. This statistical-based model did not define why nitrate decreased.

Box 2.

Example: If the amount of nitrate-N in the soil is 50 kg NO_3–N ha^{-1}, and if the first-order rate constant is 0.1 g NO_3–N (g nitrate-N × day)$^{-1}$ what is the half-life?
First-order kinetics

$$t_{half\text{-}life} = \frac{\ln 2}{k} = \frac{0.693}{0.1} = 6.93 \text{ d}$$

Example: If the amount of urea in the soil is 100, 90, 81, 73, 66, 60, and 55 mg kg^{-1} at 0, 1, 2, 3, 4, 5, and 6 d, respectively, what is the first-order rate constant for urease enzyme?
Plot time vs. natural log of remaining urea. The resulting regression equation is $y = 4.598 - 0.1003t$, $r^2 = 0.999$. The half-life is 6.91 d.

Box 3.

Example: Determine the first-order rate equation for the release of N from a slow release fertilizer incubated in water at 25°C (modified from Wang et al., 2011)

Percentage released	Amount released	Amount remaining	Natural log of amount remaining	Days
10	0.43	3.87	1.353254507	4.5
20	0.86	3.44	1.235471471	14.5
30	1.29	3.01	1.101940079	31.2
40	1.72	2.58	0.947789399	73.6
50	2.15	2.15	0.765467842	136.1
60	2.58	1.72	0.542324291	198.5

Solution: This problem can be solved by determining the regression equation between time (t) and natural log of amount remaining.

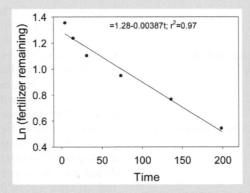

Based on this equation, the release of the N contained with the slow-release fertilizer granule has a first-order rate constant of 0.00387 g (g d^{-1}). Based on this constant, the half-life for fertilizer release is 179 d, and the amount remaining can be calculated with the equation $y_t = y_0 e^{-kt}$. The release is then calculated by subtracting remaining from the applied. The predicted release of the N can be adjusted by multiplying the release in any time period by multiplying the predicted release by a reduction factor ranging from 0 to 1. This reduction factor may by a function of temperature or soil moisture content. For example, at field capacity, the factor may be 1, whereas at the permanent wilting point the factor may be 0.01. Before using the model to predict nutrient release, it should be tested using an independent data set. This testing process is called validation.

Using Models to Solve Problems

Seeding a Cover Crop

Cover crops are being proposed as a techniques to increase nutrient retention and reduce nitrate leaching into groundwater (Reese et al., 2014). However, one of the barriers limiting cover crop success is the use of pre-emergence herbicides such as atrazine. The ability to successfully seed a cover crop depends on length of time for the herbicide to break down or the selection of the cover crop mixture. Herbicide degradation has been described using first-order kinetics and its degradation rate is determined by plotting the natural logarithm of the herbicide concentration in the soil solution vs. time (Walker, 1974; Charnay et al., 2005). The rate constant is the absolute value of the slope. The dissipation rates constants (k) for atrazine, isoproturon, and metamitron have been reported to be

0.022, 0.0533, and 0.041 g (g × day)$^{-1}$, respectively (Charnay et al., 2005). Based on these rate constants, the half-lives for atrazine, isoproturon, and metamitron are 31.5, 13.0, and 16.9 d, respectively (Box 4).

Assessing Changes in Soil Health: Carbon Turnover

Many simulation models rely on first-order kinetics to describe the conversion of fresh plant biomass to CO_2 or soil organic C (Clay et al., 2006). Simulation models are available to describe C turnover using a single decreasing exponential equation to multiple exponential equations (Clay et al., 2006). Most models require a rate constant to calculate the transformation of fresh biomass and soil organic C to CO_2. These constants can be determined using a variety of approaches including (i) placing a known amount of crop residues (nonharvested C or NHC) into residue bag, which in turn can be left on the soil surface or buried in the soil; and (ii) killing the plants and then monitoring changes in amount of above- or belowground residue with time (Chang et al., 2016). The rate constant is then determined by plotting the natural logarithm (ln) of the amount of residue remaining against time. As described above, the absolute value of the slope is the first-order rate constant. A nonharvested C value of 0.178 g (g × year)$^{-1}$ indicates that 17.8% of the C is converted to SOC in a year. The rate constants may vary with soil depth (Clay et al., 2015). In a grassland system, Chang et al. (2016) showed that the first-order rate constants for roots contained in the 0- to 15- and 15- to 30-cm soil depth were 0.00514 and 0.00205 g (g × day)$^{-1}$. A rate constant of 0.00514 g (g × day)$^{-1}$ indicates that 0.514% of the roots decompose in 1 d, and after 135 d, one-half of the roots will be decomposed (Box 5).

Determining Soil Sampling Protocols

Conceptual models by themselves have value. Clay et al. (1997) built a conceptual model that helped explain the distribution of nitrate remaining in the soil after a fertilizer application. This model hypothesized that when N was band applied, the movement of nitrate from the band with percolating water resulted in a contour map that resembles an upside down Christmas tress (Fig. 1). However, when broadcast, nitrate moved only in vertical direction (Clay et al., 1997). They assessed the ability of different soil sampling protocols by randomly

Box 4.

Example: If atrazine is applied at a rate of 2 pound acre^{-1}, how much of the herbicide remains in the soil solution after 30, 60, and 90 d if the first-order dissipation rate is 0.022 g (g × day)$^{-1}$.

Answer: Use the equation, $y_t = y_0 e^{-k}$, to solve this problem.

30 d:
$y_t =$ (2 pound acre^{-1})e$^{-0.022 \cdot 30}$ = 1.03 lb ac^{-1}

60 d:
$y_t =$ (2 pound acre^{-1})e$^{-0.022 \cdot 60}$ = 0.534 lb ac^{-1}

90 d:
$y_t =$ (2 pound acre^{-1})e$^{-0.022 \cdot 90}$ = 0.276 lb ac^{-1}

If the atrazine residual activity needs to be decreased to 0.3 pound acre^{-1}, when should it be safe to plant the cover crop?
Solve the equation, $y_t = y_0 e^{-k}$, time t.

Time = (ln 2 − ln 0.3)/0.022 = 86 d.

Box 5.

Example: If 5000 kg of surface residue are applied to the soil, and after 7, 14, 21, and 28 d 4340, 3768, 3272, and 2840 kg are remaining, what is the first-order rate constant in g (g × day)$^{-1}$ and half live?

Answer: Linear relationship between time and natural log of remaining substrate

$$k = 0.0202 \text{ g (g × day)}^{-1}$$

The half-life is 34 d.

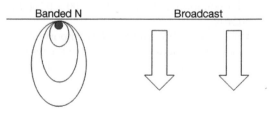

Fig. 1. Conceptualization of nitrate movement below a fertilizer band and when broadcast applied (Based on Clay et al., 1997).

sampling the two-dimensional contour map and then comparing the nutrient concentration in each core with the known amount of nitrate in the soil profile. Based on this testing, it was determined that the highest probability of obtaining a correct answer occurred when the soil core was collected half way between the row and the inter-row when N was band applied to the inter-row. In the broadcast treatment, the recommendation was to collect a random sample. Similar approaches can be used to develop effective sampling protocols for greenhouse gas emission that follow diurnal cycles (Clay et al., 1990).

Determining the Economic Optimum Fertilizer Rate

Simulation model results can be analyzed to determine the economic optimum N rate. The ease of this calculation is dependent on model used to calculate the relationship between yield and amount of fertilizer applied. In the examples below, a statistical-based polynomial model was used to simplify the calculations. The polynomial equation can be determined using many different approaches. In this example, the following data was entered into Excel:

Yield	N rate	(N rate)^2
145	75	5,625
154	100	10,000
160	125	15,625
163	150	22,500

After entering the data into Excel, select data, data analysis, and regression. If you do not have the data analysis program on your Excel, you will need to load the Analysis Tool Pak. Once regression is selected, enter $A$2:$A$6 into the Input Y range and $B$2:$C$5 into the input X range. The resulting second-order polynomial equation will be, yield = 100 + 0.78(N^1) − 0.0024(N^2).

When using this model, it is important to ensure that the model is not biased in the region where economic optimum rates are likely to occur. If it is biased, the data will either be either below or above the predicted line. This problem can be overcome by dropping out data points at the two extremes. For example, drop the lowest or the highest N rate.

Solving for the Economic Optimum Rate

The economically optimum N rate is defined as the point on the yield response curve where

$$d(\$yield) = d(\$N)$$

or where the incremental change in the corn value resulting from an increasing N rate [d($yield)] is equal to the incremental change in the cost of the fertilizer resulting from an increasing N rate [d($N)]. By definition,

$$d(\$yield) = d(yield) \times \$yield$$

and

$$d(\$N) = d(N) \times \$N$$

where, d(yield) is the incremental yield increase resulting from a higher N rate, $yield is the selling price of the grain, d($N) is the change in cost of the fertilizer, d(N) is the incremental change in the amount of N applied, and $N is the selling price of the N fertilizer. At the optimum N rate, the incremental value increase in yield is equal to the incremental increase in fertilizer cost or d($yield) = d($N). This equation was rearranged into

$$\frac{d(\$yield)\$}{d(\$N)\$} = 1$$

After substituting

$$\$yield = (yield/area) \times (\$/yield)$$

$$\$N = (N_{applied}/area) \times (\$/N)$$

into

$$\frac{d(\$yield)\$}{d(\$N)\$} = 1$$

the equation,

$$\frac{d(\$yield)\$}{d(\$N)\$} = \frac{d\left(\dfrac{yield}{area} \times \dfrac{\$}{yield}\right)}{d\left(\dfrac{N_{applied}}{area} \times \dfrac{\$}{N}\right)} = \frac{\dfrac{\$}{yield} d\left(\dfrac{yield}{area}\right)}{\dfrac{\$}{N} d\left(\dfrac{N_{applied}}{area}\right)}$$

was derived. This equation was simplified through the following steps.

$$=\frac{\dfrac{\$}{yield}d\left(\dfrac{yield}{area}\right)}{\dfrac{\$}{N}d\left(\dfrac{N_{applied}}{area}\right)}=\frac{\dfrac{\$}{yield}d(yield)}{\dfrac{\$}{N}d(N)}=1$$

and

$$=\frac{\dfrac{\$}{yield}d(yield)}{\dfrac{\$}{N}d(N)}=1$$

to

$$\frac{d(yield)}{d(N)}=\frac{\dfrac{\$}{N}}{\dfrac{\$}{yield}}=\frac{\$N}{\$yield} \qquad [1]$$

This simplification indicates that at the economic optimum value, the incremental change in yield [d(yield)] resulting from an incremental change in N fertilizer [d(N)] applied is related to the cost of the fertilizer ($N) and the crop selling price ($yield).

If yield is defined by the second-order polynomial equation,

$$y = a + b(N^1) + c(N^2)$$

then the derivative [d(yield)]/[d(N)] of yield with respect to N is

$$\frac{d(yield)}{d(N)} = b + 2cN \qquad [2]$$

The rules for this derivation include that:

1. a is deleted because it is a constant,

2. Reducing the power by one, N^1 becomes N^{1-1} or 1, and N^2 becomes N^{2-1} or N, and

3. Taking the power and multiplying it by the coefficient of the variable, cN^2 becomes $2cN$

By combining Eq. [1] and [2], the follow equations were derived:

$$\frac{d(yield)}{d(N)} = \frac{\$N}{\$yield} = b + 2cN$$

$$\frac{\$N}{\$yield} = b + 2cN$$

The economically optimum N rate is

$$N = \frac{\dfrac{\$M}{\$yield} - b}{2c} \qquad [3]$$

where the b and c terms used in this solution are based on the equation, $y = a + bN^1 + cN^2$. The examples in Boxes 6–8 use this equation to determine the economic optimum fertilizer rate.

In summary, models can be used for many different purposes. Conceptual models can be used to design effective experiments and identify basic assumptions. The way the problem is visualized can affect both the experimental design and interpretation. In addition, this chapter provides examples for (i) how to calculate zero- and first-order rate constants, (ii) how to use rate constants to solve problems, and (iii) how to use empirical relationships to calculate economic optimum rates.

Box 6.

Example: Using Eq. [3], what is the economic optimum N rate for the equation, $b = 0.78$ and $c = (-0.0024)$, if corn is selling for $3 bushel^{-1} and N costs $0.55 pound^{-1}. The economically optimum N rate is derived by substituting the appropriate values into the equation:

$$\frac{\dfrac{0.55}{3} - 0.78}{2(-0.0024)} = 124 \text{ lb N acre}^{-1}$$

Box 7.

Example: If the price of P is $60 pound^{-1}, corn is selling for $6 per bushel, and the relationship between yield and rate is yield = $100 + 1.03x - 0.00732x^2$, what is the economically optimum P rate?

Using the generalized solution for second-order polynomial equations with b being equal to 1.03 and c being equal to -0.00732, the producer should apply 63.5 pound P acre^{-1}.

$$\frac{\dfrac{\$P}{\$yield} - b}{2c} = \frac{(0.6/6) - 1.03}{2(-0.00732)} = \frac{63.5 \text{ pound P}}{\text{acre}}$$

Box 8.

Example: Using Eq. [3], what is the economic optimum P rate for the following data?

P rate)	Corn yield
pound acre^{-1}	bushel acre^{-1}
0	100
20	120
40	130
60	135
80	137

What is the second-order polynomial model between P and yield, and what is the economic optimum P rate if P sells for \$0.40 pound^{-1} and corn is sold at \$6 bushel^{-1}.

Answer: Use regression analysis to determine the relationship between the P rate and yield. Based on this analysis, the equation rating yield and P is yield $= 100 + 1.03x - 0.00732x^2$. By replacing b with 1.04 and c with -0.00732, the economic optimum P rate is 65.8 pounds of P per acre.

$$\frac{\frac{\$P}{\$yield} - b}{2c} = \frac{\left(\frac{0.4}{6}\right) - 1.03}{2(-0.00732)} = \frac{65.8 \text{ pound P}}{\text{acre}}$$

References

Adair, C.E., W.J. Parton, S.J. Del Grosso, W.L. Silver, M.E. Harmon, S.A. Hall, I.C. Burkes, and S.C. Hart. 2008. Simple three pool model accurately described patter of long-term decomposition in diverse climates. Glob. Change Biol. 14:2636–2660.

Boyle, M., and E.A. Paul. 1989. Carbon and nitrogen mineralization kinetics in soil previously amended with sewage sludge. Soil Sci. Soc. Am. J. 53:99–103. doi:10.2136/sssaj1989.03615995005300010018x

Camargo, G.G.T., and A.R. Kemanian. 2016. Six crop models differ in their simulation of water uptake. Agric. For. Meteorol. 220:116–129. doi:10.1016/j.agrformet.2016.01.013

Chang, L., D.E. Clay, A. Smart, and S. Clay. 2016. Estimating annual root decomposition in grassland systems. Rangeland Ecol. Manag. doi:10.1016/j.rama.2016.02.002 (in press)

Charnay, M.P., S. Tuis, Y. Coquet, and E. Barriuso. 2005. Spatial variability in herbicide degradation in surface and subsurface soils. Pest Manag. Sci. 61:845–855. doi:10.1002/ps.1092

Clay, D.E., C.G. Carlson, K. Brix-Davis, J. Oolman, and B. Berg. 1997. Soil sampling strategies for estimating residual nitrogen. J. Prod. Agric. 10:446–451. doi:10.2134/jpa1997.0446

Clay, D.E., C.G. Carlson, S.A. Clay, V. Owens, T.E. Schumacher, and F. Mamani Pati. 2010. Biomass estimation approach impacts on calculated SOC maintenance requirements and associated mineralization rate constants. J. Environ. Qual. 39:784–790. doi:10.2134/jeq2009.0321

Clay, D.E., C.G. Carlson, S.A. Clay, C. Reese, Z. Liu, and M.M. Ellsbury. 2006. Theoretical derivation of new stable and non-isotopic approaches for assessing soil organic C turnover. Agron. J. 98:443–450. doi:10.2134/agronj2005.0066

Clay, D.E., S.A. Clay, C.G. Carlson, and S. Murrell. 2012. Mathematics and calculations for agronomists and soil scientists: Metric version. International Plant Nutrition Institute. Peachtree Corners, GA.

Clay, D.E., G.L. Malzer, and J.L. Anderson. 1990. Ammonia volatilization from urea as influenced by soil temperature, soil water content, and nitrification and hydrolysis inhibitors. Soil Sci. Soc. Am. J. 54:263–266. doi:10.2136/sssaj1990.03615995005400010042x

Clay, D.E., G. Reicks, C.G. Carlson, J. Miller, J.J. Stone, and S.A. Clay. 2015. Tillage and corn residue harvesting impacts surface and subsurface carbon sequestration. J. Enron. Qual. 44:803–809 doi:10.2134/jeq2014.07.0322

Clay, S.A., D.E. Clay, D. Horvath, J. Pullis, C.G. Carlson, S. Hansen, and G. Reicks. 2009. Corn (*Zea mays*) responses to competition: Growth alteration vs. limiting factor. Agron. J. 101:1522–1529. doi:10.2134/agronj2008.0213x

Cramer, W., D.W. Kicklighter, A. Bondeau, B. Moore Iii, G. Churkina, B. Nemry. A. Ruimy, A.L. Schloss, and The Participants of the Potsdam NPP Model Intercomparison. 1999. Comparing global models of terrestrial net primary productivity (NPP): Overview and key results. Glob. Change Biol. 5:1–15. doi:10.1046/j.1365-2486.1999.00009.x

Halpin, T., and T. Morgan. 2008. Information and modeling and relational data bases. Elsevier, New York.

Hansen, S., H.E. Jensen, N.E. Nielsen, and H. Svendson. 1991. Simulation of the nitrogen dynamics and biomass productivity in winter wheat using the Danish simulation model DAISY. Fert. Res. 27:245–259. doi:10.1007/BF01051131

Kemanian, A.R., and C.O. Stöckle. 2010. C-Farm: A simple model to estimate the carbon balance of soil profiles. Eur. J. Agron. 32:22–29. doi:10.1016/j.eja.2009.08.003

Kim, K.I., D.E. Clay, S. Clay, and C.G. Carlson. 2008. Do synergistic relationships between nitrogen and water influence the ability of corn to use nitrogen derived from fertilizer and soil? Agron. J. 100:551–555. doi:10.2134/agronj2007.0064

Kim, K.I., D.E. Clay, S. Clay, G.C. Carlson, and T. Trooien. 2013. Testing corn (*Zea Mays* L.) regional nitrogen recommendation models in South Dakota. Agron. J. 105:1619–1625. doi:10.2134/agronj2013.0166

Lehmann, J., and M. Kleber. 2015. The contentious nature of soil organic matter. Nature 528:60–68. doi:10.1038/nature16069

MathWorks. 2015. MATLAB and statistics toolbox release. The MathWorks, Inc., Natick, MA.

Moriles, J., S. Hansen, D.P. Horvath, G. Reicks, D.E. Clay, and S.A. Clay. 2012. Microarray and growth analyses identify differences and similarities of early corn response to weeds, shade, and nitrogen stress. Weed Sci. 60:158–166. doi:10.1614/WS-D-11-00090.1

Nira, R., and A. Nishimune. 1993. Studies on nitrogen mineralization properties of Tokachi soils by kinetic analysis. Soil Sci. Plant Nutr. 39:321–329. doi:10.1080/00380768.1993.10417003

O'Neill, P.M., J.F. Shanahan, J.S. Schepers, and B. Caldwell. 2004. Agronomic responses of corn hybrids from different eras to deficit and adequate levels of water and nitrogen. Agron. J. 96:1660–1667. doi:10.2134/agronj2004.1660

Parton, W.J., J.M.O. Scurlock, D.S. Ojima, T.G. Gilmanov, R.J. Scholes, D.S. Schimel, et al. 1993. Observations and modeling of biomass and soil organic matter dynamics for the grassland biome worldwide. Global Biogeochem. Cycles 7:785–810. doi:10.1029/93GB02042

Reese, C., D.E. Clay, S.A. Clay, A. Bich, A. Kennedy, S. Hansen, and J. Miller. 2014. Wintercover crops impact on corn production in semi-arid regions. Agron. J. 106:1479–1488. doi:10.2134/agronj13.0540

Setiyono, T.D., H. Yang, D.T. Walters, A. Dobermann, R.B. Ferguson, D.F. Roberts, D.J. Lyon, D.E. Clay, and K.G. Cassman. 2011. Maize-N: A decision tool for nitrogen management in maize. Agron. J. 103:1276–1283. doi:10.2134/agronj2011.0053

Tebaldi, C., K. Hayloe, J.M. Arblaster, and G.A. Meehl. 2006. Going to the extremes: Am intercomparison of model simulated historical and future extreme events. Clim. Change 79:185–211. doi:10.1007/s10584-006-9051-4

Walker, A. 1974. Simulation model for predicting herbicide persistence. J. Environ. Qual. 3:396–401. doi:10.2134/jeq1974.00472425000300040021x

Wang, S., A.K. Alva, Y. Li, and M. Zhang. 2011. A rapid technique to predicting the nutrient release from polymer coated controlled release fertilizer. Open J. Soil Sci. 1:40–44. doi:10.4236/ojss.2011.12005

Optimization of Financially Constrained Fertilizer Use

Charles Wortmann* and Kayuki Kaizzi

Smallholder farmers often use little fertilizer for crop production because of financial constraints. Demands for use of available money is high and returns on investment in fertilizer use need to be very high for use to be competitive. Potential profitability varies greatly with choice of nutrient, the crop to which the nutrient is applied, and the rate of application. This chapter addresses a fertilizer use optimization approach intended to maximize net returns for the farmer's economic and agronomic context. It discusses relevant basic principles and describes an approach applied in Uganda. Reliable crop–nutrient response functions for an agroecological zone are essential for estimating likely profitability of crop–nutrient rate decisions. These functions can be determined using results from locally conducted research and extrapolation of results from other locations with similar growing conditions. Smallholder farms often produce four or more substantial crops and optimization of fertilizer use may involve 12 or more crop–nutrient combinations to determine the optimal rate for a farmer's situation. A computer-based fertilizer optimization tool and its paper version are described.

Many smallholder farmers live in ongoing financial peril with little opportunity for improvement and much vulnerability. Investment in fertilizer use competes with other uses of financial resources for meeting immediate needs. Therefore, fertilizer investments must give high returns with little risk. Fertilizer use often costs two to six times as much in sub-Saharan Africa compared with the United States or Europe because of high importation and transportation costs and inefficient markets as a result, in part, to low volume of sales (Vlek, 1990; Sanchez, 2002). Bank loans for fertilizer use are commonly lacking and fertilizer use subsidies, where they exist, often are of relatively little benefit to the more financially constrained and, too often, subsidize less-effective fertilizer types (Heisey and Mwangi, 1996). These high fertilizer costs are often coupled with low and highly variable commodity values.

Abbreviations: AEZ, agroecological zone; AfSIS, Africa Soil Information System; DAP, diammonium phosphate; EOR, mean economically optimal rate; FOT, fertilizer optimization tool; GYGA, Global Yield Gap Atlas; MRR, marginal rates of return; TSP, triple superphosphate.

Charles Wortmann, Dep. of Agronomy and Horticulture, Univ. Nebraska–Lincoln, 279 Plant Science, Lincoln, NE 68583-0915. *Corresponding author (cwortmann2@unl.edu). Kayuki Kaizzi, National Agricultural Research Laboratory, PO Box 7065, Kampala, Uganda (kckaizzi@gmail.com).

doi:10.2134/soilfertility.2014.0088

Soil Fertility Management in Agroecosystems
Amitava Chatterjee and David Clay, Editors

Smallholder farmers often have limited financial ability to use fertilizer but need high returns on their small investment. Achieving high returns commonly reduces future financial constraints, enabling farmers to invest more in fertilizer use in following years. A guideline used in evaluating the potential for adoption of practices by financially constrained farmers is that there be a 100% net return on investment within a year, otherwise an alternate use for the capital is more likely to result in a better return (CIMMYT, 1988; Wortmann and Ssali, 2001). However, even this high rate of return can be unattractive to those who need to use their financial resources for meeting short-term needs or to make longer-term investments as for education of their children.

Optimization of fertilizer use by smallholder farmers refers, in this chapter, to the maximization of net returns on the farmers' investment, while it may refer to maximizing productivity or preventing environmental problems or soil degradation in some other contexts. Excessive fertilizer costs, severe financial constraints, variable commodity values, and risks associated with crop production require smallholders to make wise fertilizer use decisions according to their financial resources. The choice of crop–nutrient rate combinations is very important to maximizing net returns.

Crop response to applied nutrients is commonly curvilinear until yield reaches a plateau, recognizing that yield sometimes does decrease from a peak with high application rates. Such yield declines from the peak can be due to localized fertilizer salt effects, severe soil water depletion at a critical growth stage following a period of vigorous growth, increased disease, or more lodging. Once a crop–nutrient response curve is determined that represents the typical

response of that crop for the targeted growing conditions, mean monetary returns on investment in nutrient application can be estimated (Sawyer and Nafziger, 2005). A curvilinear to plateau response function of corn (*Zea mays* L.) to applied N in Uganda is represented in Fig. 1. The effect on grain yield is great with low N application rates, but there is a diminishing rate of return with decreasing rate of yield increase per unit of N as the maximum yield is approached until no more yield increase occurs. Such a crop–nutrient response can be represented by an asymptotic function such as $Y = a - b \times c^N$, where Y is yield, a is yield at the plateau, b is the possible yield increase resulting from nutrient application, c is a curvature coefficient giving an abrupt response at low c values and having a value of 1 with a linear response, and N is the nutrient application rate. The curvilinear to plateau response function can also be captured by other functions such as an exponential rise to a maximum (modified Mitscherlich) or a spherical (linear to nonlinear rise to plateau) function (Dobermann et al., 2011). Maize response to applied N in Uganda was represented by $Y = 3.711 - 1.823 \times 0.96^N$ (Fig. 1; Kaizzi et al., 2012c).

Crop–Nutrient Response Functions and Maximizing Net Returns to Fertilizer Use

The fertilizer nutrient application rate that on average maximizes net return to fertilizer use is often called the mean economically optimal rate (EOR; Dobermann et al., 2011) or rate of maximum return (Sawyer and Nafziger, 2005). Determination of this rate needs to consider

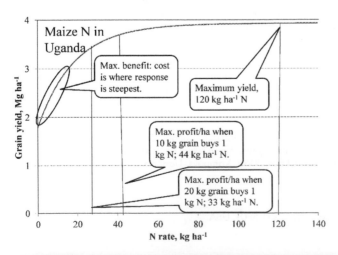

Fig. 1. Crop–nutrient response functions can be used to determine the nutrient rate required to approach maximum yield, maximum profit per hectare as a result of nutrient application for different nutrient costs relative to commodity value, and high returns on small investments (adapted from Jansen et al., 2013).

the costs of using the fertilizer nutrient including purchase, transport, and application costs. It also needs to consider the expected value of the commodity to the farmer. These costs and values can be expressed as amount of grain needed to pay for the use of each unit of fertilizer nutrient (kg kg^{-1}). For N applied to maize in Uganda, an average of 120 kg ha^{-1} N was required to achieve 98% of maximum yield (Fig. 1). However, when the cost of using 1 kg of fertilizer N is equal to the farmgate value of 10 or 20 kg of maize, the EOR for N is 44 or 33 kg ha^{-1} N, respectively. Where fertilizer can be used without financial constraints, fertilizer use is optimized when net return per hectare is maximized, above which net returns to fertilizer use diminish while yield increases until it reaches the maximum. In December 2013, for example, it was determined for western Uganda that 14 kg of maize were needed to cover the costs of each kilogram of N use and the mean EOR was 40 kg ha^{-1} N. Many poor farmers, however, do not have enough money to apply fertilizer at EOR for all of their land. They need to maximize returns on their small investment, which is achieved by applying at a rate where the curve of yield response to applied nutrient is steep. However, the potential profitability of a particular crop–nutrient option needs to be considered relative to other application options.

Some Crop–Nutrient Responses Offer Much More Profit Potential than Others.

Individual crop–nutrient response functions for a targeted production situation are important for maximizing net profit to fertilizer use. However, some crop–nutrient responses offer much more profit potential than others. This is illustrated by results from research in Uganda (Kaizzi et al. (2012a,b,c, 2014; Jansen et al., 2013) in Fig. 2 showing

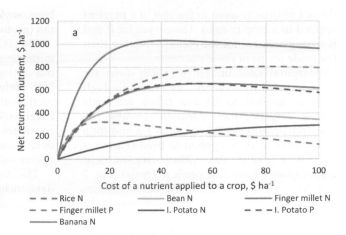

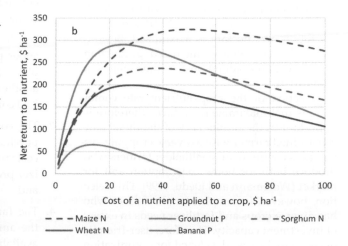

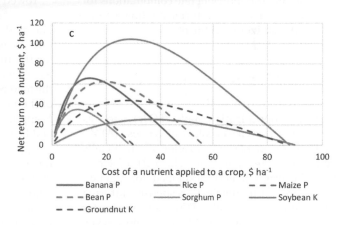

Fig. 2. Crop–nutrient options differ in their profit potential from (a) most profitable to (c) least profitable. The profit potential is affected by the crop–nutrient response function as well as fertilizer cost and commodity value.

the net returns (*y*-axis) to the cost of a nutrient applied to a crop (*x*-axis). The curves for P and K were generated for the elemental rather than the oxide forms of these nutrients.

Marginal rates of return (MRR) differ for crop–nutrient options and are especially high with N applied to banana (*Musa acuminata* Colla), finger millet [*Eleusine coracana* (L.) Gaertn.], upland rice (*Oryza sativa* L.), and for a small amount of N applied to bean (*Phaseolus vulgaris* L.) (Fig. 2a). Other crop–nutrient combinations have less profit potential (Fig. 2 b,c). As investment in a crop nutrient increases, the slope of the MRR curve decreases to become parallel to the slopes of MRR for other crop–nutrient combinations. Where the crop–nutrient slopes are parallel, MRR are similar. The nutrient investment rate at which a MRR curve reaches a peak is where MRR is zero and net return per hectare is maximized. Beyond the peak, the value of increased production is less than the added cost of more nutrient application.

Selection of the crop–nutrient rate combinations that maximize net returns therefore is important to the profitability of the financially constrained farmer with a diverse cropping system as is common for smallholder farmers who produce both for home consumption and for market (Wortmann and Eledu, 1999). This selection, however, is complex for the farmer who has diverse crops and fertilizer needs in excess of investment capacity. Therefore, user-friendly decision tools were developed for optimization of choice of crop–nutrient rate combinations for maximization of net returns on finance-constrained fertilizer investment.

The Fertilizer Optimization Tool

Fertilizer optimization tools (FOTs) such as developed for agroecological zones (AEZs) of Uganda enable optimization of financially constrained fertilizer use. The FOT of Fig. 3 was the original FOT developed for Uganda. As more crop–nutrient response information became available, FOTs specificity for AEZs were improved and there are now seven FOTs for major agricultural AEZs of Uganda with different crops and response functions. Crops currently served by AEZ specific FOTs in Uganda include maize, sorghum [*Sorghum bicolor* (L.) Moench], upland rice, wheat (*Triticum aestivum* L.), finger millet, Irish potato (*Solanum tuberosum* L. subsp. *tuberosum*), banana,

bean, soybean [*Glycine max* (L.) Merr.], and peanut (groundnut; (*Arachis hypogaea* L.) (Fig. 2).

The FOTs use linear optimization to determine, on average, the most profitable fertilizer use option specific for a farmer's context. The FOT optimizes solutions using the Solver add-on (Frontline Systems, Inc.) of Microsoft Office Excel 2007 or later. The process stage of the FOT considers the farmer-specified constraints, predetermined model constraints, and the model's optimization mode (Fig. 4).

The farmer-imposed constraints, or input data, include the following (Fig. 3):

1. The intended land area to be planted and the expected commodity value at harvest for each crop to be planted, where zero is entered for land area of crops that are not being considered;

2. Fertilizers available including urea, triple super phosphate (TSP), diammonium phosphate (DAP), muriate of potash (KCl), or another available product with its N-P$_2$O$_5$–K$_2$O content specified;

3. The cost of using each fertilizer including purchase, delivery, application, and interest costs with zero entered for the cost if a fertilizer product is not of interest to the farmer; and

4. The farmer's budget constraint defined as the amount of money that the farmer has available for fertilizer use, whether borrowed or cash on hand.

The model is constrained by setting maximum rates to avoid exceeding the range of inference for the underlying equations such as in cases where fertilizer is free or of very low cost. As the field research results supported Liebig's law of the minimum, the FOT often requires some N application before P can be applied to cereals and bean and some P application before K can be applied. This is not always the case, such as for banana, as N and K application gave similar yield responses and greater responses than with P. The FOTs do not consider expected yield or rainfall amount, soil test results, or previous crop, as these did not have significant effects on the crop–nutrient responses (Kaizzi et al. 2012a,b,c); this is further addressed below.

The FOT uses Microsoft Office Excel Solver to reiteratively find a solution that optimizes a specified mathematical function, often referred to as an objective function, subject to the specified constraints. The objective function in this

Producer Name:	xxx
Prepared By:	xxx
Date Prepared:	June 19, 2012

Fig. 3. An image of a fertilizer optimization tool for Uganda showing the data (a) input and (b) output screens (adapted from Jansen et al., 2013).

Crop Selection and Prices		
Crop	Area Planted (Ha)*	Expected Grain Value/kg †
Maize	0	0
Sorghum	0	0
Upland rice, paddy	0	0
Beans	0	0
Soybeans	0	0
Groundnuts, unshelled	0	0
Total hectares	0	

Fertilizer Selection and Prices				
Fertilizer Product	N	P2O5	K2O	Price/50 kg bag ¶*
Urea	46%	0%	0%	0
Triple super phosphate, TSP	0%	46%	0%	0
Diammonium phosphate, DAP	18%	46%	0%	0
Murate of potash, KCL	0%	0%	60%	0
xxx	%	%	%	0

b

Budget Constraint	
Amount available to invest in fertilizer	0

Fertilizer Optimization					
	Application Rate - kg/Ha				
Crop	Urea	TSP	DAP	KCL	xxx
Maize	0	0	0	0	0
Sorghum	0	0	0	0	0
Upland rice, paddy	0	0	0	0	0
Beans	0	0	0	0	0
Soybeans	0	0	0	0	0
Groundnuts, unshelled	0	0	0	0	0
Total	0	0	0	0	0

Expected Average Effects per Ha		
Crop	Yield Increases	Net Returns
Maize	0	0
Sorghum	0	0
Upland rice, paddy	0	0
Beans	0	0
Soybeans	0	0
Groundnuts, unshelled	0	0

Total Expected Net Returns to Fertilizer	
Total net returns to investment in fertilizer	0

case is to maximize net returns to fertilizer use as the difference of added crop revenue and added fertilizer costs, subject to the constraints imposed by the farmer or internal to the FOT. The crop–nutrient response functions are combined with expected crop values and fertilizer use costs to select the optimal crop–nutrient rate combinations that deliver the highest net return on investment within the financial constraint. This is calculated by relating to a circular reference, where each combination must satisfy all constraints imposed by the user and FOT.

Once the optimal crop–nutrient rate combinations have been determined, the results are displayed (Fig. 3b) including the optimized crop–fertilizer application rates and the expected mean fertilizer effects on yield and

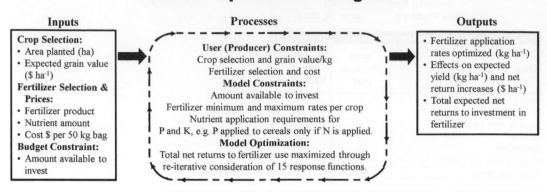

Fertilizer Optimization Diagram

Inputs	Processes	Outputs
Crop Selection: • Area planted (ha) • Expected grain value ($ ha⁻¹) **Fertilizer Selection & Prices:** • Fertilizer product • Nutrient amount • Cost $ per 50 kg bag **Budget Constraint:** • Amount available to invest	**User (Producer) Constraints:** Crop selection and grain value/kg Fertilizer selection and cost **Model Constraints:** Amount available to invest Fertilizer minimum and maximum rates per crop Nutrient application requirements for P and K, e.g. P applied to cereals only if N is applied. **Model Optimization:** Total net returns to fertilizer use maximized through re-iterative consideration of 15 response functions.	• Fertilizer application rates optimized (kg ha⁻¹) • Effects on expected yield (kg ha⁻¹) and net return increases ($ ha⁻¹) • Total expected net returns to investment in fertilizer

Fig. 4. Diagram for the Uganda Fertilizer Optimization Tool (adapted from Jansen et al., 2013).

net returns to fertilizer use for each crop. Output data also include the total quantity of each fertilizer needed and the expected mean overall net profit. Each set of constraints imposed by the user delivers a unique but optimized solution based on attributes relevant to the farmer's production and economic context.

What if the Farmer or Farmer Advisor Does Not Have a Computer?

Well-functioning computers are often scarce in smallholder farming communities. Therefore, paper FOT versions have been developed. Table 1 shows the current paper FOT for the <1400-m elevation AEZ of eastern Uganda. It makes assumptions of fertilizer type availability, fertilizer costs, and commodity values. It also provides guidelines for method and time of calibrated application; therefore, assumptions are made for readily available fertilizer measurement units and for crop row spacing.

The paper FOT considers three levels of farmer financial ability and fertilizer use guidelines are provided based on MRR. Financial ability Level 1 represents the most financially constrained and Level 3 represents no financial constraint to fertilizer use. Therefore, the MRR needs to be >$4 $⁻¹ for a financial Level 1 choice, that is a profit of $4 ha⁻¹ for each additional $1 ha⁻¹ of nutrient applied. Choices for Level 1 financial ability in this FOT include some N and P applied to finger millet, N applied to bean, and P applied to groundnut (Table 1). If the farmer has one or more of these crops in the cropping system, these fertilizer application options have priority over other options.

For financial ability Level 2, the MRR is between 2 and 4. Some nutrient applied to five different crops qualifies and with increased rates compared with financial ability Level 1 (Table 1). The farmer may not have the financial ability to apply fertilizer for all crops, but the five choices offered have somewhat similar profit potential. Therefore, the farmer can opt for choices according to financial ability but ensuring that Level 1 options are addressed. For financial ability Level 3, the MRR is between 0.1 and 2, and the rates are at or very near mean EOR.

The paper version FOT does need to be updated when there are substantial changes in the costs of fertilizer use relative to commodity values. Also, about 20% of the profit potential in the decision making is lost with the paper FOT than the Excel FOT because of the generalization for farmer's budget constraint and selection a set choices with financial capability levels.

How Do Fertilizer Optimization Decisions Account for Alternative Practices?

The FOTs operate on a near whole-farm basis, in that several crops likely grown across several land parcels or fields are considered. The fields differ in soil productivity and in management history. For example, one or more may have had manure application recently or there may be a residual nutrient contribution from a previous application of manure. Another may have had a leguminous green manure or cover crop, or a field may have a maize–bean intercrop. Soil test values, if available, likely differ. Rather than attempting to

Table 1. Example of a paper version of an agroecological-zone-specific fertilizer optimization tool.

UGANDA (eastern low-altitude) fertilizer use optimizer: paper version

Assumptions:

· Measurement is with a water bottle cap that holds ~5 g of urea, 6 g of diammonium phosphate (DAP), and 8 g of triple superphosphate (TSP) or KCl, or with a bottle cut at 2 cm to hold 50 g of urea, 60 g of DAP, or 80 g of TSP or KCl.

· Maize and sorghum have 75-cm row spacing; finger millet, bean, soybean, and groundnut have 50-cm row spacing.

· Fertilizer costs per 50-kg bags are Uganda shillings 120,000 for urea, 120,000 for TSP, and 100,000 for KCl.

· Commodity values per kilogram are Uganda shillings 350 for maize, 300 for sorghum, 1200 for finger millet, 1500 for bean, 800 for soybean, and 2000 for groundnut (unshelled).

Level 1 financial ability

· Finger millet: Apply 27 kg ha^{-1} urea and 30 kg ha^{-1} TSP at planting time (apply in a band with a bottle lid, urea for 3.7 m and TSP for 5.6 m). Sidedress 27 kg ha^{-1} urea (apply in a band with a bottle lid for 3.7 m)

· Bean: Apply 40 kg ha^{-1} urea at planting time (apply in a band with a bottle lid for urea for 2.5 m)

· Groundnut: Apply 64 kg ha^{-1} TSP at planting time (apply in a band with a bottle lid for 2.6 m).

Level 2 financial ability

· Maize: Apply 25 kg ha^{-1} urea at planting time (apply in a band with a bottle lid for 2.7 m). Sidedress 25 kg ha^{-1} urea (apply in a band with a bottle lid for 2.7 m).

· Sorghum: Apply 27 kg ha^{-1} urea at planting time (apply in a band with a bottle lid for 2.5 m)

· Finger millet: Apply 42 kg ha^{-1} urea and 42 kg ha^{-1} TSP at planting time (apply in a band with a bottle lid, urea for 2.4 m and TSP for 4.0 m). Sidedress 42 kg ha^{-1} urea (apply in a band with a bottle lid for 2.4 m)

· Bean: Apply 49 kg ha^{-1} urea at planting time (apply in a band with a bottle lid for 2.0 m)

· Groundnut: Apply 35 kg ha^{-1} TSP at planting time (apply in a band with a bottle lid for 4.8 m).

Level 3 financial ability (maximizes profit per hectare)

· Maize: Apply 25 kg ha^{-1} urea at planting time (apply in a band with a bottle lid for 2.7 m) Sidedress 25 kg ha^{-1} urea (apply in a band with a bottle lid for 2.7 m).

· Sorghum: Apply 32 kg ha^{-1} urea at planting time (apply in a band with a bottle lid for 2.1 m). Sidedress 32 kg ha^{-1} urea (apply in a band with a bottle lid for 2.1 m).

· Finger millet: Apply 67 kg ha^{-1} urea and 67 kg ha^{-1} TSP at planting time (apply in a band with a bottle lid, urea for 1.5 m and TSP for 2.5 m). Sidedress 67 kg ha^{-1} urea (apply in a band with a bottle lid for 1.5 m)

· Bean: Apply 67 kg ha^{-1} urea and 72 kg ha^{-1} TSP at planting time (apply in a band with a bottle lid, urea for 1.5 m and TSP for 2.3 m)

· Groundnut: Apply 131 kg ha^{-1} TSP and 54 kg ha^{-1} KCl at planting time (apply in a band with a bottle lid, TSP for 1.3 m and KCl for 2.9 m).

· Soybean: Apply 72 kg ha^{-1} TSP at planting time (apply in a band with a bottle lid for 2.3 m).

address all such factors in the already complex FOT, these are considered in a second step of the decision process to adjust the fertilizer rates determined by the FOT (Table 2). For example, for each 1 t ha^{-1} of farmyard manure applied, application rates of urea, triple super phosphate, and KCl can be reduced by 5, 3, and 2 kg ha^{-1}, respectively. Cereal–bean intercropping calls for more fertilizer P but no change in fertilizer N compared with that determined for the cereal sole crop. If soil test P is >15 mg kg^{-1}, fertilizer P is not applied but if soil test K is <100 mg kg^{-1}, added K should be applied.

What Is the Data Source for Determination of Crop–Nutrient Response Functions?

Data availability for determination of crop–nutrient response functions where smallholder agriculture prevails is often inadequate. The relatively good availability of such data for Uganda is the exception (Kaizzi et al. 2012a,b,c, 2013, 2014). Therefore, a data set of results from a large number of trials conducted in sub-Saharan Africa since the 1970s has been compiled and, where appropriate, the results reanalyzed

Table 2. Guidelines to determine optimal fertilizer rates in consideration of other practices or soil test information specific to a field.

ISFM practice	Fertilizer reduction			
	Urea	DAP or TSP†	KCl	NPK 17:17:17
Previous crop was a green manure crop (%)	100	70	70	70
	kg ha⁻¹			
Fresh vegetative material (e.g. prunings of *Lantana* or *Tithonia*) applied, per tonne of fresh material	9.9	4.9	4.9	19.8
Farmyard manure per tonne of dry material	12.4	7.4	4.9	24.7
Residual value of farmyard manure applied for the previous crop, per tonne	4.9	2.5	2.5	7.4
Dairy or poultry manure, per tonne dry material	22.2	9.9	12.4	39.5
Residual value of dairy and poultry manure previously applied for the previous crop, per tonne	4.9	4.9	2.5	7.4
Compost, per tonne	19.8	7.4	7.4	37.1
Residual value of previously applied compost per tonne	7.4	4.9	2.5	12.4
Rotation	0% reduction but more yield expected			
Cereal–bean intercropping	Increase DAP/TSP by 17 kg ha⁻¹, but no change in N and K compared with sole cereal fertilizer			
Cereal–other legume (effective in N fixation) intercropping	Increase DAP/TSP by 27 kg ha⁻¹, reduce urea by 22 kg ha⁻¹, and no change in K compared with sole cereal fertilizer			
If Mehlich III P > 15 mg kg⁻¹	Apply no P			
If soil test K < 100 mg kg⁻¹	Band apply 50 kg ha⁻¹ KCl			

† DAP, diammonium phosphate; TSP, triple superphosphate.

according to the asymptotic curvilinear to plateau response function defined above $Y = a - b \times c^N$. This resulted in >4000 georeferenced crop–nutrient response functions across 14 crops, although with many for maize and very few for some other crops. The database is growing as results from current research are added.

Spatial data availability has greatly improved for Africa in recent years. Soil property layers are available at 1-km grid resolution through the Africa Soil Information System (AfSIS) (http://africasoils.net/). Climate property layers are available such as through Global Yield Gap Atlas (GYGA) (http://yieldgap.org). The crop response database together with available spatial information gives opportunity to extrapolate information and find results from growing conditions similar to conditions of the targeted AEZ. For some crops in Uganda, such as Irish potato where response information was lacking, response functions were determined by query of information from locations of similar growing conditions as the targeted AEZ. The environmental window of the query was defined using the aridity index and seasonality spacial layers of GYGA (van Wart et al., 2013); the soil organic C, sand, and pH layers of AfSIS; the absolute value of latitude; and elevation. The query boundaries were as follows:

Aridity index: ± 1000 to 6000; if >6000, omit <5000

Seasonality: ± 1000

Latitude (absolute value): ± 5°

Elevation: if >1000 m, ± 300 m; if <1000 m, omit >1300 m

Soil organic C: ±10 if <35 g kg⁻¹; if >35, omit <25

Sand: ±20 if >75%; if <70%, omit >80%

pH: if <5.4, then ±0.4; if pH >5.4, omit pH <5.0

While priority should be given to locally generated research results, extrapolated results can well complement the local results.

Conclusions

Smallholder farmers often are severely constrained financially and are unable to invest much in fertilizer use. Investments in fertilizer use need to give very high net returns to compete with other demands for the use of available money. Achieving high returns can lead to reduced poverty constraints and can gradually increase fertilizer use and crop productivity. Reliable crop–nutrient response functions for an AEZ are important to determine the mean profit potential of applying a nutrient to a crop. Once response functions are available for several or all important crops of the AEZ, the crop–nutrient rate combinations that will maximize mean net profit from fertilizer use can be determined. Consideration of the many variables affecting fertilizer use optimization is best done with a decision tool, that is, a FOT. This approach has been developed for nine important food crops of Uganda with FOTs specific for seven AEZs. Both electronic and paper versions of the FOT are developed. The paper versions are important because of computer scarcity in smallholder communities, although some potential profit efficiency is lost, and the paper FOTs need to be revised when substantial changes occur in cost of fertilizer use relative to commodity values. In cases of inadequate response information for an AEZ, information extrapolated from other locations of similar growing conditions was applied.

References

CIMMYT. 1988. From agronomic data to farmer recommendations: An economics training manual. CIMMYT, Mexico City.

Dobermann, A., C.S. Wortmann, R.B. Ferguson, G.W. Hergert, C.A. Shapiro, D.D. Tarkalson, and D. Walters. 2011. Nitrogen Response and Economics for Irrigated Corn in Nebraska. Agron. J. 103:67–75. doi:10.2134/agronj2010.0179

Heisey, P.W., and W. Mwangi. 1996. Fertilizer use and maize production in sub-Saharan Africa. Econ. Work. Pap. 96-01. CIMMYT, Mexico City.

Jansen, J., C.S. Wortmann, M.C. Stockton, and K.C. Kaizzi. 2013. Maximizing net returns to financially constrained fertilizer use. Agron. J. 105:573–578. doi:10.2134/agronj2012.0413

Kaizzi, C.K., J. Byalebeka, O. Semalulu, I. Alou, W. Zimwanguyizza, A. Nansamba, P. Musinguzi, P. Ebanyat, T. Hyuha, and C.S. Wortmann. 2012a. Sorghum response to fertilizer and nitrogen use efficiency in Uganda. Agron. J. 104:83–90. doi:10.2134/agronj2011.0182

Kaizzi, C.K., J. Byalebeka, O. Semalulu, I. Alou, W. Zimwanguyizza, A. Nansamba, P. Musinguzi, P. Ebanyat, T. Hyuha, and C.S. Wortmann. 2012b. Maize response to fertilizer and nitrogen use efficiency in Uganda. Agron. J. 104:73–82. doi:10.2134/agronj2011.0181

Kaizzi, C.K., J. Byalebeka, O. Semalulu, I. Alou, W. Zimwanguyizza, A. Nansamba, E. Odama, and C.S. Wortmann. 2014. Upland rice response to fertilizer in Uganda. Afr. J. Plant Sci. 8:416–425. doi:10.5897/AJPS2014.1175

Kaizzi, C.K., C. Wortmann, J. Byalebeka, O. Semalulu, I. Alou, W. Zimwanguyizza, A. Nansamba, P. Musinguzi, P. Ebanyat, and T. Hyuha. 2012c. Optimizing smallholder returns to fertilizer use: Bean, soybean and groundnut. Field Crops Res. 127:109–119. doi:10.1016/j.fcr.2011.11.010

Kaizzi, C.K., C. Wortmann, and J. Jansen. 2013. More profitable fertilizer use for poor farmers. Better Crops Plant Food 97(3):4–6.

Sanchez, P.A. 2002. Soil fertility and hunger in Africa. Science 295:2019–2020. doi:10.1126/science.1065256

Sawyer, J.E., and E.D. Nafziger. 2005. Regional approach to making nitrogen rate decisions for corn. In: Proc. Thirty-Fifth North Central Extension-Industry Soil Fertility Conf., Des Moines, IA. 16–17 Nov. 2005. Vol. 21. Potash and Phosphate Inst., Brookings, SD. p. 16–24.

van Wart, J., L.G.J. van Bussel, J. Wolf, R. Licker, P. Grassini, A. Nelson, H. Boogaard, J. Gerber, N.D. Mueller, L. Claessens, M.K. van Ittersum, and K.G. Cassman. 2013. Use of agro-climatic zones to upscale simulated crop yield potential. Field Crops Res. 143:44–55. doi:10.1016/j.fcr.2012.11.023

Vlek, P.L.G. 1990. The role of fertilizers in sustaining agriculture in sub-Saharan Africa. Fert. Res. 26:327–339. doi:10.1007/BF01048771

Wortmann, C.S., and C.A. Eledu. 1999. Uganda's agroecological zones: A guide for planners and policy makers. CIAT, Kampala, Uganda.

Wortmann, C.S., and H. Ssali. 2001. Integrated nutrient management for resource-poor farming systems: A case study of adaptive research and technology dissemination in Uganda. Am. J. Altern. Agric. 16:161–167. doi:10.1017/S0889189300009140

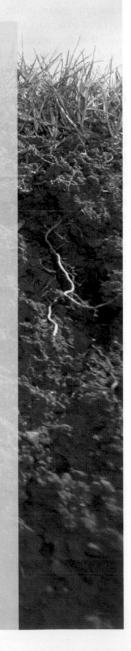

- Cover Crops as Nitrogen Scavengers and Nitrous Oxide Emissions

- Fertilizer Replacement Value

- Protection against Phosphorus Loss through Soil Erosion

- Soil Organic Matter Improvement

- Cover Crop Species

- Cover Crop Management Effect on Soil Fertility

- Barriers of Cover Crops Adoption and Research Needs

Cover Crops Impacts on Nitrogen Scavenging, Nitrous Oxide Emissions, Nitrogen Fertilizer Replacement, Erosion, and Soil Health

Amitava Chatterjee* and David E. Clay

By reducing nutrient leaching and erosional losses, increasing soil organic matter, fixing N_2, and providing a living mulch and forage for livestock, cover crops can improve long-term economic environmental sustainability. However, obtaining these benefits depends on the selection of the most appropriate plant species and management practices. For example, annual ryegrass (*Lolium multiforum* Lam.) can develop dense, shallow root systems within a short period, creating a vegetative mat that provides a cushioning effect against rain drop impact and helps prevent erosion. Other species, such as barley (*Hordeum vulgare* L.), may have very deep roots, which are efficient in scavenging nutrients. Leguminous crops, like hairy vetch (*Vicia villosa* Roth), and clover (*Trifolium* spp.), have the ability to fix atmospheric N and release it to subsequent crops. The objectives of this chapter are to: (i) provide an overview of how cover crops may fit into a soil fertility program, (ii) discuss the use of cover crops as N scavengers and the impact on N_2O emissions, (iii) provide examples of how to determine fertilizer replacement value, (iv) discuss the impact of cover crops on P and soil loss through erosion, and (v) examine cover crop impact on soil health.

By reducing nutrient losses, trapping soil nutrients, and fixing N, the inclusion of cover crops (Table 1) in the rotation can increase the nutrient efficiency of the entire system. Understanding the benefits and weaknesses of the interactions among management, soil, and climatic conditions is critical for optimizing cover crop systems. There are many different types of cover crop seeding options available, including: planting the cover crop prior to planting the cash crop in the spring, planting cover crops in-season during the summer, or planting a cover crop in the fall following harvest (Smeltekop et al., 2002; Bich et al., 2014;

Abbreviations: EONR, economic optimum N rate; NFRV, nitrogen fertilizer replacement.

A. Chatterjee, Dep. of Soil Science, North Dakota State University, Fargo, ND 58108-6050; D.E. Clay, Soil Science, South Dakota State University, Brookings, SD 57007 (David.Clay@sdstate.edu). *Corresponding author (amitava.chatterjee@ndsu.edu).

doi:10.2134/soilfertility.2016.0012

Soil Fertility Management in Agroecosystems
Amitava Chatterjee and David Clay, Editors

Table 1. Common cover crop species and their role in soil fertility management (Clark, 2007).

Cover crop species	Scientific name	Type	Characteristics related to soil fertility and health
Nonlegume			
Annual ryegrass	*Lolium multiflorum* Lam.	Cool season annual grass	· Dense root system improves water movement · High N user
Buckwheat	*Fagopyrum esculentum* Moench	Summer or cool-season annual broadleaf grain	· Abundant and fine roots, release mild acid that solubilize fertilizer · Takes up P
Rye	*Secale cereale* L.	Cool season annual cereal grain	· Best cool season for N scavenging, as much as 28 to 56 kg N ha^{-1} with fibrous root system · Takes up deep soil K and increase surface soil exchangeable K
Sorghum–sudangrass hybrids	*Sorghum bicolor* × *S. bicolor* var. *sudanese*	Summer annual grass	· Mowing increases root mass to penetrate deeper and reduce subsoil compaction.
Barley	*Hordeum vulgare* L.	Cool season annual cereal grain	· Overwintering barley root grows as deep as 2 m. · Holds soil strongly with fibrous root system and reduces erosion control · Scavenges nutrients
Oats	*Avena sativa* L.	Cool season annual cereal	· Takes up excess N and small amounts of P and K
Legumes			
Berseem clover	*Trifolium alexandrium* L.	Summer annual or winter annual legume	· Dry matter N concentration is about 2.5%, and single cut can supply 56 to 112 kg aboveground N ha^{-1}.
Crimson clover	*Trifolium incarnatum* L.	Winter annual or summer annual legume	· N contribution ranges from 78–168 kg N ha^{-1}.
Field pea	*Pisum sativum* subsp. *arvense* (L.) Asch.	Summer annual and winter annual legume	· Austrian winter pea is top N producer, ranging from 100 to 336 kg N ha^{-1}. · Most moisture-efficient crop in producing biomass
Medics	*Medicago* spp.	Winter annual or summer annual legume	· 3.5–4% of tissue N contributing 224 kg N ha^{-1} · Low dense vegetation breaks raindrop impact, and root extends up to 1.5 m. · It can establish in summer drought prone areas
Red clover	*Trifolium pratense* L.	Short-lived perennial, biennial, or winter annual legume	· Full-season overwintered plants can fix 78–168 kg N ha^{-1}. Wide adaptability, excellent soil conditioner with an extensive root system
Hairy vetch	*Vicia villosa* Roth	Winter annual or summer annual legume	· Heavy contributor of mineralized N · Improves root zone water recharge over winter by reducing runoff · More drought tolerant than other vetches · Higher plant P concentrations than crimson or red clover
Sweet clovers, yellow and white	*Melilotus officinalis* (L.) Pall. *M. alba* Medik.	Biennial, summer annual, or winter annual legume	· Ability to extract P and K from insoluble minerals · Vesicular–arbuscular mycorrhizal fungi associated with roots increase P availability · N fixing potential 112 to 325 kg N ha^{-1}

Fisher et al., 2014). Depending on climatic, tillage, and soil conditions, each option has unique strengths and weaknesses and can increase, not impact, or reduce yield (Abdollahi and Munkholm, 2014; Reese et al., 2014; Baas et al., 2015). For example, a cover crop planted in the spring provides early season soil protection, but may directly compete with the cash crop for nutrients (Smeltekop et al., 2002), while a cover crop planted following fall harvest may have little opportunity for growth and development, especially under cold and dry conditions. In semiarid environments, cover crops can replace fallow, but the cover crop's water use may result in reduced yields of the following cash crop (Army and Hide, 1959; Reese et al., 2014). In humid, warm environments when water is not a limiting factor, cover crops can increase yields and improve soil health and nutrient recycling. In semiarid frigid regions, yields can be increased by trapping snow, building soil organic matter, and increasing cropping diversity (Fig. 1). In addition, decreased available water can reduce the plants' ability to respond to abiotic and biotic stresses (Hansen et al., 2013; Reese et al., 2014; Poeplau and Don, 2015). When considering a cover crop, the entire system must be evaluated because the benefits and weaknesses can extend beyond the duration of the cover crop.

Cover Crops as Nitrogen Scavengers and Nitrous Oxide Emissions

Fall and winter cover crops can improve N efficiency by reducing $NO_3–N$ leaching losses during the fall, winter, and spring (Acuña and Villamil, 2014; Reese et al., 2014). Both legume and nonlegume cover crops scavenge N by transforming inorganic N into organic N, which in turn, reduces N losses through denitrification and leaching (Dabney et al., 2007; Reese et al., 2014). According to Wagger and Mengel (1988), nonleguminous cover crops during their growth cycle: (i) reduce the supply of soil inorganic nitrogen; (ii) release N during the mineralization of cover crop biomass; (iii) immobilize N during the decomposition of the cover crop biomass; and (iv) reduce N losses. Nitrogen mineralized during decomposition has the potential to reduce the following crop's N requirement. A synchrony between cover crop N release and main crop N demand is critical to avoid yield reductions and optimize fertilizer efficiency (Salmerón et al., 2010).

Cover crops can have a mixed impact on greenhouse gas emissions from soil. Nitrous oxide emissions may be reduced by decreasing soil NO_3^- concentrations and soil wetness. Increased N_2O emission may also occur if root exudates stimulate microbial metabolism and deplete the oxygen supply (Drury et al., 1991). Jarecki et al. (2009) reported that in central Iowa, rye (*Secale cereale* L.) cover crop reduced N_2O emissions from manure treated soils. Mitchell et al. (2013) reported that a rye cover crop in Iowa reduced N_2O emission when no N was applied but increased emissions when the fertilizer N at the rate of 152 kg N ha^{-1} was applied. Basche et al. (2014) reviewed 26 papers that provided 106 observations on the impact of cover crops on N_2O emissions. They reported that 40% of the observations indicated that cover crops decreased emissions; factors like fertilizer N rate, soil N incorporation, and rainfall had control on N_2O emissions. Moreover, legume cover crops produced higher N_2O emissions at lower

Scavenging Nutrients	Reduce Erosion Loss	Improve Soil Organic Matter	Fixing N
•Reduce nutrient leaching •Extensive root system access nutrients hard to reach by main crop •Bring nutrients like, phosphorus, potassium and calcium to surface from deep profile	•Reduce run off water velocity •Reduce raindrop impact •Stabilize soil particles with root system	•Deep roots can open up the clay pan •Decaying roots creates channels for root growth •Reduce compaction through root cusion •Enhances mycorrhizal activity •Encourage soil aggregation	•Leguminous cover crops fix atmospheric N •Mineralization of cover crop residues supply N to main crop

Fig. 1. Role of cover crops in maintaining soil fertility.

fertilizer N rate and vice versa, whereas non-legume cover crops increased N_2O emissions with increasing fertilizer N rate.

Subsurface tile drainage enhances the downward movement of NO_3^-. This sometimes leads to more NO_3^- leaching into drainage lines and hence, greater offsite transport. Cover crops showed a wide variation in their ability to reduce NO_3^- leaching, ranging from 6 to 94% of the amount leached from no cover crop areas (Kaspar et al., 2007; Kaspar and Singer, 2011; Drury et al., 2014). In the north-central United States, Strock et al. (2004) reported that a winter rye cover crop in a corn (*Zea mays* L.) followed by soybean [*Glycine max* (L.) Merr.] system reduced subsurface tile drainage total discharge by 11% and the amount of NO_3^- discharge by 13%. Kaspar et al. (2007) also found that winter rye significantly reduced subsurface drainage water flow–weighted NO_3^- concentrations by 59% and NO_3^- loads by 61%. Salmerón et al. (2010) reported that barley and winter rape (*Brassica napus* L.) treatments reduced NO_3^- leaching by 80% compared to no cover crop treatments mainly due to reduction of NO_3^- concentration in drainage water. Drury et al. (2014) found that cover crops reduced the 5-yr flow-weighted mean NO_3^- concentration in tile drainage water ranging from 21 to 38% and cumulative NO_3^- loss by 15% relative to no cover crop treatment. Cover crop and controlled tile drainage subirrigation together reduced the flow-weighted mean NO_3^- concentration up to 47% relative to no cover crop and unrestricted tile drainage. Cover crop reduced runoff and simultaneously increased tile drainage, primarily due to increases in near-surface soil hydraulic conductivity. Drury et al. (2014) concluded that the integration of cover crop and the water-table management system was highly effective for reducing NO_3^- loss from cool, humid agricultural soils.

Fertilizer Replacement Value

Cover crop growth can result in a reduction in the amount of fertilizers required to optimize yields through at least two mechanisms. The first mechanism is that legume cover crops can fix atmospheric N, which can help build the amount of soil N. However, the benefits for N fixation may be dependent on the amount of N in soil and nodulation. Smeltekop et al. (2002) reported that when soil N was low, annual medic (*Medicago* spp.) can fix N, which subsequently is made available to corn as the cover crop senesced.

However, when soil N was high after fertilizer application, medic utilized the N fertilizer, which produced N stress in corn. A second method of fertilizer N reduction is observed when N taken up by the cover crop is made available to the cash crop through mineralization. The change in the N fertilizer recommendation due to cover crop presence is called the nitrogen fertilizer replacement value (Fig. 2, Box 1). The N fertilizer replacement value of a winter cover crop is the difference between the amount of N required to obtain the economically optimum yield in rotations with and without the cover crop. For example, if 100 kg N ha^{-1} are required to produce the economically optimum 5 Mg ha^{-1} corn yield in a rotation with a cover crop and 130 kg N ha^{-1} are required to produce the economically optimum yield 4.8 mg ha^{-1}, then the replacement value of the cover crop is 30 kg N ha^{-1}. For reliable estimates, the nitrogen fertilizer replacement value (NFRV) should be calculated based on fertilizer N response curves for the two rotations (Ketterings et al., 2015). Ketterings et al. (2015) indicated that NFRV can be calculated using two basic approaches, traditional and difference. In the traditional approach, NFRV is based on the amount of fertilizer required to achieve a given yield without the cover crop. For example, if the yield following the cover crop without any N additions is 10 Mg ha^{-1} and to achieve this yield without the cover crop requires 40 kg N ha^{-1}, then the replacement value is 40 kg N ha^{-1}. In this approach, the experiment should contain two cover crop treatments (cover crop and no cover crop) and multiple N rates that are targeted to the rate where the replacement power is likely to occur. For example, if it is expected that the N replacement is 40 kg N ha^{-1}, then the experiment should contain rates surrounding 40 kg N ha^{-1}. This calculation does not consider whether the cover crop impacted the yield potential.

In the difference approach, the economic optimum N rates between the cover crop and no cover crop treatments are compared. In this experiment, the N rates should bracket the expected economic optimum N rate. One of the strengths of this calculation is that it will assess whether the cover crop changed the yield potential. The economic optimum N rate is the relationship between fertilizer inputs and yield. There are many ways to define this relationship, but the problem can be simplified if the relationship is defined with a second-order polynomial equation [yield = a + b(inputs) + c(inputs)^2]. If this model is used, the economic optimum N rate (EONR) is defined by the equation:

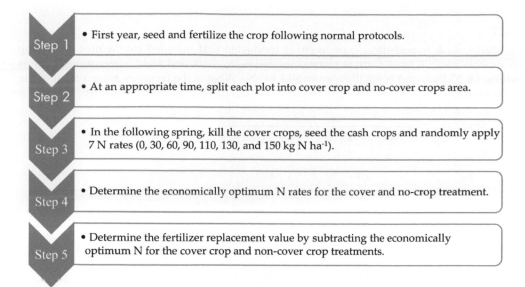

Step 1
• First year, seed and fertilize the crop following normal protocols.

Step 2
• At an appropriate time, split each plot into cover crop and no-cover crops area.

Step 3
• In the following spring, kill the cover crops, seed the cash crops and randomly apply 7 N rates (0, 30, 60, 90, 110, 130, and 150 kg N ha⁻¹).

Step 4
• Determine the economically optimum N rates for the cover and no-crop treatment.

Step 5
• Determine the fertilizer replacement value by subtracting the economically optimum N for the cover crop and non-cover crop treatments.

Fig. 2. An experiment for calculating fertilizer replacement value in a winter cover crop experiment followed by corn.

Box 1.

Problem 1. What is the cover crop N credit based on the traditional approach (Ketterings et al., 2015) in the following experiment? In this experiment, the N response is measured in areas with and without cover crops.

Fertilizer N	No cover crop	Cover crop
kg ha⁻¹	Mg ha⁻¹	Mg ha⁻¹
0	5	7
30	7	9
60	9	10
90	10	11
120	11	11

Answer: This problem can be analyzed using two basic approaches, traditional and difference. Comparing yields for the 0 and 30 kg N ha⁻¹ treatments shows that yields in the no cover crop–30 kg N h⁻¹ and cover crop–0 kg N ha⁻¹ treatments were identical. Based on this comparison, it is determined that the N replacement credit was 30 kg N ha⁻¹.

$$\text{EONR} = \frac{\dfrac{\text{cost of inputs}}{\text{value of products}} - b}{2c}$$

A comparison of both methods shows that there is value to both approaches. However, slightly different experiments are used for both methods. For the traditional approach (Box 1, Problem 1), the experiment should focus on the area surrounding the likely fertilizer replacement value, whereas in the difference method (Box 2, Problem 2) the experiment should focus on the area surrounding the EONR. These two methods will produce different answers, and when discussing the values, the method must be clearly defined. An alternative step-by-step guide for estimating the release of N from cover crops is available

Box 2.

Problem 2. A hypothetical experiment is conducted that included two winter cover crop treatments and six rates. Corn is planted the following year, and yield is measured. Corn is selling for $5 bu^{-1}, and N fertilizer costs $0.4 lb^{-1} N. What is the fertilizer replacement of the cover crop?

N rate	Corn yield	
	No cover crop	Cover-crop
lb acre^{-1}	———— bu acre^{-1} ————	
40	120	130
60	160	190
90	190	200
120	210	220
150	215	225
200	215	225

Answer:

Step 1. Determine the second-order polynomial equations for both systems.

No cover crop: yield = 52.1 + 2.071 (N rate) – 0.00639 (N rate)^2

Cover crop: yield = 74.3+1.969 (N rate) – 0.00616 (N rate)^2

Step 2. Calculate EONR

$$\text{No cover crop EONR} = \frac{\frac{0.4}{5} - 2.071}{2 \times (-0.00639)} \qquad \text{Cover crop EONR} = \frac{\frac{0.4}{5} - 1.969}{2 \times (-0.00616)}$$

$$= 99 \text{ lb acre}^{-1} \qquad\qquad\qquad = 95 \text{ lb acre}^{-1}$$

Based on this calculation, the EONR is 4 lb N acre^{-1} (99 – 95). This calculation does not consider the change in the yield potential. The 10 bu acre^{-1} higher yield is achieved with less N.

in Sullivan and Andrews (2012). When using the second-order polynomial equation to define the N yield response, it may be biased in region surrounding the EONR. If the model is biased in this region, the EONR may be inaccurate. Under these conditions, the accuracy of the calculation can be improved by fitting the data to a second-order polynomial plateau model as opposed to a second-order polynomial model (Cerrato and Blackmer, 1990). The second-order polynomial and second-order polynomial models are defined by the equations:

Second-order polynomial: yield = $a + b$(inputs) + c(inputs)^2

Second-order polynomial plateau model:

Yield = $a + b$(inputs) + c(inputs)^2, where yield < maximum yield

Yield = maximum yield, where inputs > maximum yield

In the second-order polynomial plateau model, the curves are continuous and they meet at a value where the slopes are identical. Based on the model, the N rate where the yield reaches a maximum ($dy/dN = 0$) is calculated with the equation $-b/2c$. However, these two models may still be biased. This problem is solved by using iteration to recalculate the a, b, and c values (Box 3).

Legume cover crops have typical NFRV values ranging 0 to 100 kg N ha^{-1} (Reeves 1994; Mahama et al., 2016). Wagger (1989) estimated corn recovery of legume N was between 40 and

Box 3.

Problem 3. What are the intersections of the second-order polynomial and plateau models for the following equations?

No cover crop: yield = 52.1 +2.071 (N rate) − 0.00639 (N rate)^2

Cover crop: yield = 74.3+1.969 (N rate) − 0.00616 (N rate)

Answer:

$$\text{No cover crop:} \frac{-b}{2c} = \frac{-2.071}{2 \times -0.00639} = 162$$

$$\text{Cover crop:} \frac{-1.969}{2 \times 0.00616} = 159 \text{ lb N acre}^{-1}$$

45 kg N ha^{-1}, which represented about 36% of the total N content of crimson clover (*Trifolium incarnatum* L.) and 30% of the total N content of hairy vetch. Moreover, Miguez and Bollero (2005) reported that grass winter cover crops had no effect on corn yield, whereas legume winter cover crops increased corn yield by 37% without any N fertilizer additions, but this positive effect was reduced when N fertilizer was applied. In the northern US Corn Belt, however, there may not be adequate time between harvest and frost for cover crop seed germination and establishment (Singer, 2008).

Protection against Phosphorus Loss through Soil Erosion

Unlike N that is lost through leaching in the soil profile, P is immobile in soil and held so tightly that it does not leach, unless attached to leaching sediments or organic compounds (Uusitalo et al., 2001). A major mechanism of P off-site movement is with surface sediments through runoff and wind erosion. Cover crops provide soil cover that protects the soil against these erosional forces. Cover crops reduce erosional losses by (i) absorbing raindrop impact, (ii) reducing aggregate detachment, (iii) increasing surface roughness, (iv) delaying runoff initiation, (v) intercepting runoff, (vi) reducing runoff velocity, (vii) increasing time for water infiltration, (viii) reducing soil nutrient concentrations in the soil solution, and (ix) influencing compaction (Acuña and Villamil, 2014; Blanco-Canqui et al., 2015). By reducing the nutrient concentration load in runoff water, cover crops reduce nonpoint-source pollution. Cover crops also improve the soil physical environment by

improving aggregation and reducing compaction. Abdollahi and Munkholm (2014) reported that 5 yr of using a cover crop reduced plow pan compaction at the 20- to 40-cm depth as measured by reducing penetration resistance. This, in turn, can increase water infiltration and reduce the propensity for water erosion and gully formation. Reduction in runoff loss mainly protects particulate P loss with variable response was observed in the case of soluble P loss (Kaspar and Singer, 2011). Soluble P concentration of runoff can be estimated by the following equation (Sharpley, 1985):

$$P_r = \frac{KP_a DBt\alpha W\beta}{V}$$

where P_r is the average soluble P concentration of an individual runoff event (mg L^{-1}), P_a is the available P content (Bray-1) in the surface 50 mm of soil (mg kg^{-1}) before each runoff event, D is the depth of interaction between surface soil and runoff (mm), B is the bulk density (Mg m^{-3}), t is runoff event duration (minutes), W is runoff during the event (m), and K, α, β are constants for a given soil. Sharpley and Smith (1991) used this equation to predict the impact of a cover crop on the soluble P concentrations in runoff water. These calculations indicate that cover crops had a minimal impact on soluble P movement; however, cover crop following peanut (*Arachis hypogaea* L.) reduced runoff, soil loss, and total P loss (Table 2). Sharpley and Smith (1991) concluded that soluble P concentrations in runoff might increase with cover crops as a function of soil fertility, crop type, and growth stage and that cover crops can improve the P uptake of succeeding crops by absorbing relatively unavailable native and residual fertilizer P and release labile P to succeeding crops. In

Table 2. The impact of cover crops on runoff, soil loss, soluble P movement, and total P loss. This experiment was based on data collected from a peanut experiment, and soil loss between October and April at Fort Cobb, OK (modified from Sharpley and Smith, 1991).

	1985		1988	
	No cover crop	Cover crop (rye)	No cover crop	Cover crop (wheat)
Runoff, cm	2.9	1.12	12.22	2.54
Soil loss, kg ha^{-1}	4612	998	17,858	2206
Soluble P movement, μg g^{-1}	0.14	0.19	0.12	0.15
Total P, kg ha^{-1}	1.35	0.47	5.88	0.92

addition, decomposition of cover crops releases CO_2 into the soil, which facilitates P dissolution from primary minerals by changing pH in soil microsites (Fageria et al., 2005). Growing cover crops, particularly legumes, increases the abundance of mycorrhizal fungi and mycorrhizal hyphae trap nutrients inaccessible by most plant roots. Mycorrhizal relationships with roots also reduce nutrient leaching loss (Dabney et al., 2007).

Soil Organic Matter Improvement

Cover crops add biomass residues that ultimately may contribute to the soil organic matter (Reicosky and Forcella, 1998; Ding et al., 2006). However, the benefit of cover crop species on soil organic matter may be dependent on the crop species. Cover crop residue with low C/N ratio (<20:1) will mineralize the nutrients faster than a cover crop mix with a high C/N ratio (>40:1) (Chatterjee, 2013). Cover crop benefits are more evident under no-tillage due to rapid decomposition of residues as compared

to conventional-tillage management systems. Cover crops, by providing a surface mulch, can reduce soil temperature diurnal fluctuations (Dabney et al., 2007). Sainju et al. (2002) found that integrating cover crop, no-tillage, and fertilization increased organic C and N concentrations, but a nonlegume cover crop (rye) was better than legumes (hairy vetch and crimson clover) and N fertilization in increasing concentrations of soil organic C and N. Kuo and Sainju (1998) used first-order kinetics [$N_{mineralization} = N_o (1 - e^{-kt})$] to explain N mineralization from cover crop mixtures (rye, vetch, and ryegrass) with different C/N ratios. In this greenhouse study, the rate constants (k) for rye and vetch were 0.075 and 0.143 wk^{-1}. Mixtures were between these extremes and vetch had a half-life of 4.9 wk, whereas rye had a half-life of 10.6 wk. An example of organic C addition calculation is provided in Box 4. Over the long term, cover crops provide carbon to the soil, which ultimately can enhance the soil health and productivity (Dabney et al., 2010).

Box 4.

Problem 4. Based on first-order kinetics, cover crop biomass mineralizes at a rate of 0.25 yr^{-1}. What is the half-life of the added biomass? And, if 1000 kg C are added in Year 1, how much will remain after 3 yr?

Answer:

$$\text{Half-life} = \frac{\ln 2}{k} = \frac{0.6931}{0.25} = 2.77 \text{ yr}$$

$$\text{Remaining biomass}_t = \text{biomass}_{t=0} e^{-kt}$$

$$= 1000 e^{-3 \times 0.25}$$

$$= 472 \text{ kg biomass C ha}^{-1}$$

Cover Crop Species

Table 3 presents cover crops options based on growers' objectives for different US farm regions. A survey of farmers across the central part of the United States found that popular cover crop species in Indiana and Illinois were cereal rye, wheat (*Triticum aestivum* L.), and red clover (*Trifolium pratense* L.), whereas in Iowa, cereal rye and oat (*Avena sativa* L.) were popular (Singer, 2008; Baas et al., 2015). Cover crop rooting depth varies with the interaction among species, soil compaction, soil water movement, and planting date (Dabney et al., 2007). Kristensen and Thorup-Kristensen (2004) reported that Italian ryegrass (*Lolium multflorum* Lam.), winter rye, and fodder radish (*Raphanus sativus* L. var. *oleiformis* Pers.) had rooting depths of 0.6, 1.1, and more than 2.4 m, respectively. They also reported that fodder radish had practically removed all soil NO_3^- down to the 2.5-m depth, leaving only 18 kg NO_3^-–N ha^{-1}. Chen and Weil (2010) also found that both forage radish and rapeseed had more than twice the number of roots than rye. Selection of cover crop species should consider the rooting depth, particularly in soils or climates prone to leaching, like in humid climates or sandy soils. Mixtures of two or more cover crop species often provide more benefits than single species. Compared to pure stands, mixtures of different cover crop species (e.g., cover crop "cocktails") often produce more overall biomass, tolerate adverse conditions, increase winter survival, provide more ground cover, and either scavenge more N or fix more N. High residual soil N promotes the establishment of grasses, whereas the low soil N facilitates legume growth. Kuo and Jellum (2002) reported average top growth biomass was higher for bicultural—vetch and rye or vetch and ryegrass (e.g., a legume plus grass)—than for a single species (monoculture) of grass or legume alone (e.g., rye, vetch, or ryegrass). Total N accumulation was greatest under vetch, followed by the bicultures, and lowest for monoculture of rye or ryegrass.

Cover Crop Management Effect on Soil Fertility

The impact of the cover crop on the cash crop's growth and development is influenced by many factors, including: (i) the cash crops' nutrient uptake characteristics, (ii) the C/N ratio of the cover crop, (iii) when the cover crop growth is terminated (Kuo and Sainju, 1998), and (iv) the impact of cover crop on water stress in the cash crop. If the cover crop reduces available water in semiarid and arid environments, cash crop yields can be reduced. In frigid environments, cover crops can be killed over winter by extreme cold. If the cover crop survives the

Table 3. The cover crop options and barriers to adoption for different regions in the United States (Clark, 2007).

| US farm regions | Cover crop options | | Barriers in adoption |
	N source	Soil health	
Northeast	Red clover, hairy vetch, berseem, sweet clover	Annual rye grass, sweet clover, sorgum–sudangrass hybrid, rye	Time and cost of establishment and water management
Southeast	Hairy vetch, red clover, berseem, crimson clover	Annual rye grass, rye sweet clover, sorgum–sudangrass hybrid,	Precipitation and availability and economy of irrigation
Corn Belt	Hairy vetch, red clover, berseem, crimson clover	Rye, barley, sorgum–sudangrass hybrid, sweet clover	Cash crop planting and harvest coincide with cover crop kill and planting dates
Northern Plains	medic, hairy vetch, sweet clover	Rye, barley, medic, sweet clover	Rainfall amount, cool and wet springs delay soil warming
Southern Plains	Winter peas, medic, hairy vetch	Rye, barley, medic	Water, integration into rotation and reduced effectiveness of pre-emergence herbicides in high-residue systems
Southwest	Medic, subterranean clover	Subterranean clover, medic, barley	Cooler temperatures above and below mulch
Northwest	Winter peas, hairy vetch	Medic, sweet clover, rye, barley	Rainfall amount and distribution and the irrigation availability

overwintering conditions in more temperate environments, cover crop growth can be terminated by mechanical and/or chemical techniques. The contribution of the N contained in the cover crop biomass to the cash crop may decrease or increase with delays in growth termination (Kuo and Sainju, 1998; Sainju and Singh, 2001). Early termination of the cover crop results in narrower C/N ratio in the residue, but the total residue biomass is reduced. Cover crops such as rye or barley are often terminated at a stage of development that results in wide C/N ratio (>30:1) and leads to an initial immobilization of N during the cropping season (Reeves, 1994). Komatsuzaki and Wagger (2015) reported that the cover crop planting date influenced soil inorganic N distribution at each termination date. The October planting date in North Carolina (in the southeastern United States) resulted in lower soil profile inorganic N levels than either a November or December planting date. Further, Sainju and Singh (2001) found that corn yield and N uptake was higher with late cover crop planting in no-tillage than conventional tillage practices in Georgia.

The method of termination (mechanical or herbicide) has the potential to affect the scavenging of N by the cover crop. Wortman et al. (2012) found that undercutting cover crops increased soil NO_3^- within the 0- to 20-cm depth relative to disk incorporation. In addition, undercutting increased soil moisture content compared to termination with the disk during early main crop growth. The use of herbicide on the main crop and as a termination strategy for the cover crop can restrict future cropping choices. Information on cover crop tolerances to various herbicides are available in Stahl (2016).

Barriers of Cover Crops Adoption and Research Needs

Despite multiple benefits of cover crops, the adoption rate has been relatively slow (Singer et al., 2007; Baas et al., 2015). Slow adoption rates may be related to complex interactions between management and cultural practices (Robertson et al., 2013). Singer (2008) reported that between 2001 and 2005 only 11% of Corn Belt farmers had adopted cover crops. Cover crops need to attain certain criteria for their successful adoption in crop rotations, such as (i) ease of seeding and establishment, (ii) rapid growth rate to provide ground coverage quickly, (iii)

sufficient quantity of dry matter for maintenance of residues, (iv) disease resistance while not hosting diseases of the cash crop, (v) ease of termination, and (vi) economic viability (Arbuckle and Roesch-McNally, 2015; Reeves, 1994; Baas et al., 2015). Each US farm region has additional limitations, but some are common, including (i) limited growth conditions, such as the uncertainty of having water for establishment or using too much water and drying the soil for the cash crop; (ii) narrow time frames for substantial biomass production; and (iii) a lack of suitable methods for cover crop termination (Table 3). It is evident that cover crop management, environmental conditions, and other production practices intricately link with each other in terms of cover crop potentials to increase or sustain soil fertility. The success of cover crop adoption depends on understanding these relationships and developing suitable recommendation programs more tailored to each region's challenges that are sound enough to increase the magnitude of its benefits while overcoming the limiting factors.

In summary, cover crops will impact the soil fertility. However, the impacts will be site specific and dependent on management and cover crop species. This chapter discusses several key issues associated with cover crop management including their impact on soil N, N_2O emissions, fertilizer replacement values, the use of models such as the second-order polynomial plateau model to define yield responses, erosion, and soil health.

References

Abdollahi, L., and L.J. Munkholm. 2014. Tillage system and cover crop effects on soil quality: I. Chemical, mechanical, and biological properties. Soil Sci. Soc. Am. J. 78:262–270. doi:10.2136/sssaj2013.07.0301

Arbuckle, J.G., Jr., and G. Roesch-McNally. 2015. Cover crop adoption in Iowa: The role of perceived practice characteristics. J. Soil Water Conserv. 70:418–429. doi:10.2489/jswc.70.6.418

Acuña, J.C.M., and M.B. Villamil. 2014. Short-term effects of cover crops and compaction on soil properties and soybean production in Illinois. Agron. J. 106:860–870. doi:10.2134/agronj13.0370

Army, T.J., and J.C. Hide. 1959. Effect of green manure on dryland wheat production in the Great Plains area of Montana. Agron. J. 51:196–198. doi:10.2134/agronj1959.00021962005100040004x

Baas, D., S. Casteel, D.E. Clay, S. Clay, P. Gross, J. Hoorman, R. Hoorman, T. Kaspar, E. Kladivko, M. Ruark, N. Millar, H. Murray, S. Noggle, and D. Wyse. 2015. Integrating cover crops in soybean rotations: Challenges and recommendations for the north central region. Midwest Cover Crop Council, North Central Soybean Research. Available at http://mccc.msu.edu/documents/2015Integrating_CoverCrops_Soybeans.pdf (accessed Sept. 2016).

Basche, A.D., F.E. Miguez, T.C. Kaspar, and M.J. Castellano. 2014. Do cover crops increase or decrease nitrous oxide emissions? A meta analysis. J. Soil Water Conserv. 69:471–482. doi:10.2489/jswc.69.6.471

Bich, A.D., C.L. Reese, A.C. Kennedy, D.E. Clay, and S.A. Clay. 2014. Corn yield is not reduced by mid-season establishment of cover crop in northern Great Plains environments. Crop Manage. doi:10.2134/CM-2014-0009-RS.

Blanco-Canqui, H., T.M. Shaver, J.L. Lindquist, C.A. Shapiro, R.W. Elmore, C.A. Francis, and G.W. Hergert. 2015. Cover crops and ecosystem services: Insights from studies in temperate soils. Agron. J. 107:2449–2474. doi:10.2134/agronj15.0086

Cerrato, M.E., and A.M. Blackmer. 1990. Comparison of models for describing yield response to nitrogen fertilizer. Agron. J. 82:138–143. doi:10.2134/agronj1990.00021962008200010030x

Chatterjee, A. 2013. North-central US: Introducing cover crops in the rotation. Crops and Soils 46:14–15.

Chen, G., and R.R. Weil. 2010. Penetration of cover crop roots through compacted soils. Plant Soil 331:31–43. doi:10.1007/s11104-009-0223-7

Clark, A., Editor. 2007. Managing cover crops profitably. 3rd ed. Sustainable Agricultural Research & Education (SARE). http://www.sare.org/Learning-Center/Books/Managing-Cover-Crops-Profitably-3rd-Edition (accessed 13 Oct. 2016)

Dabney, S.M., J.A. Delgado, J.J. Meisinger, H.H. Schomberg, M.A. Liebig, T. Kaspar, J. Mitchell, and D.W. Reeves. 2010. Using cover crop and cropping systems for nitrogen management. In: J.A. Delgado and R.F. Follett, editors, Advances in nitrogen management for water quality. Soil and Water Conserv. Soc., Ankeny, IA. p. 230–281.

Dabney, S.M., J.A. Delgado, and D.W. Reeves. 2007. Using winter cover crops to improve soil and water quality. Commun. Soil Sci. Plant Anal. 32:1221–1250. doi:10.1081/CSS-100104110

Ding, G., X. Liu, S. Herbert, J. Novak, D. Amarasiriwardena, and B. Xing. 2006. Effect of cover crop management on soil organic matter. Geoderma 130:229–239. doi:10.1016/j.geoderma.2005.01.019

Drury, C.F., J.A. Stone, and W.I. Findlay. 1991. Influence of cover crops on denitrification and nitrogen mineralization. In: W.L. Hargrove, editor, Cover crops for clean water. Proceedings of an international conference, West Tennessee Experiment Station, Jackson, TN. 9–11 Apr. 1991. Soil Water Con. Soc., Ankey, IA. p. 94–96

Drury, C.F., C.S. Tan, T.W. Welacky, W.D. Reynolds, T.Q. Zhang, T.O. Oloya, N.B. McLaughlin, and J.D. Gaynor. 2014. Reducing nitrate loss in tile drainage water with cover crops and water-table management systems. J. Environ. Qual. 43:587–598. doi:10.2134/jeq2012.0495

Fageria, N.K., V.C. Baligar, and B.A. Bailey. 2005. Role of cover crops in improving soil and row crop productivity. Commun. Soil Sci. Plant Anal. 36:2733–2757. doi:10.1080/00103620500303939

Fisher, B., C.K. Gerber, K.D. Johnson, E.J. Kladivko et al. 2014. Midwest cover crop field guide. 2nd ed. Purdue Extension, Lafayette, IN.

Hansen, S., S.A. Clay, D.E. Clay, C.G. Carlson, G. Reicks, J. Jarachi, and D. Horvath. 2013. Landscape features impacts on soil available water, corn biomass, and gene expression during the late vegetative growth stage. Plant Genome 6. doi:10.3835/plantgenome2012.11.0029

Jarecki, M.K., T.B. Parkin, A.S.K. Chan, T.C. Kaspar, T.B. Moorman, J.W. Singer, B.J. Kerr, J.L. Hatfield, and R. Jones. 2009. Cover crop effect on nitrous oxide emissions from manure treated Mollisol. Agric. Ecosyst. Environ. 134:29–35. doi:10.1016/j.agee.2009.05.008

Kaspar, T.C., D.B. Jaynes, T.B. Parkin, and T.B. Moorman. 2007. Rye cover crop and gamagrass strip effects on NO_3 concentration and load in tile drainage. J. Environ. Qual. 36:1503–1511. doi:10.2134/jeq2006.0468

Kaspar, T.C., and J.W. Singer. 2011. The use of cover crops to manage soil. In: J.L. Hatfield and T.J. Sauer, editors. Soil management: Building a stable base for agriculture. ASA and SSSA, Madison, WI. p. 321–337. doi:10.2136/2011.soilmanagement.c21

Ketterings, Q.M., S.N. Swink, S.W. Duiker, K.J. Czymmek, D.B. Beegle, and W.J. Cox. 2015. Integrating cover crops for nitrogen management in corn systems on Northeastern US dairies. Agron. J. 107:1365–1376.

Komatsuzaki, M., and M.G. Wagger. 2015. Nitrogen recovery by cover crops in relation to time of planting and growth termination. J. Soil Water Conserv. 70:385–398. doi:10.2489/jswc.70.6.385

Kristensen, H.L., and K. Thorup-Kristensen. 2004. Root growth and nitrate uptake of three different catch crops in deep soil layers. Soil Sci. Soc. Am. J. 68:529–537. doi:10.2136/sssaj2004.5290

Kuo, S., and E.J. Jellum. 2002. Influence of winter cover crop and residue management on soil nitrogen availability and corn. Agron. J. 94:501–508. doi:10.2134/agronj2002.5010

Kuo, S., and U.M. Sainju. 1998. Nitrogen mineralization and availability of mixed legume and non-legume cover crop residues in soil. Soil. Biol. Fertil. Soils 26:346–353. doi:10.1007/s003740050387

Mahama, G.Y., P.V. Vara Prasad, K.L. Roozeboom, J.B. Nippert, and G.W. Rice. 2016. Cover crops, fertilizer nitrogen rates, and economic return of grain sorghum. Agron. J. 108:1–16. doi:10.2134/agronj15.0135

Miguez, F.E., and G.A. Bollero. 2005. Review of corn yield response under winter cover cropping systems using meta-analytic methods. Crop Sci. 45:2318–2329. doi:10.2135/cropsci2005.0014

Mitchell, D.C., M.J. Castellano, J.E. Sawyer, and L. Pantoja. 2013. Cover crop effect on nitrous oxide emissions: Role of mineralizable carbon. Soil Sci. Soc. Am. J. 77:1765–1773. doi:10.2136/sssaj2013.02.0074

Poeplau, C., and A. Don. 2015. Carbon sequestration in agricultural soils via cultivation of cover crops—A meta-analysis. Agric. Ecosyst. Environ. 200:33–41. doi:10.1016/j.agee.2014.10.024

Reese, C., D.E. Clay, S.A. Clay, A.D. Bich, A.C. Kennedy, S.A. Hansen, and J. Moriles. 2014. Winter cover crops impact on corn production in semiarid regions. Agron. J. 106:1479–1488. doi:10.2134/agronj13.0540

Reeves, D.W. 1994. Cover crops and rotations. In: J.L. Hatfield and B.A. Steward, editors, Advances in soil science: Crops residue management. CRC Press, New York. p. 125–172.

Reicosky, D.C., and F. Forcella. 1998. Cover crop and soil quality interactions in agroecosystems. J. Soil Water Conserv. 53:224–229.

Robertson, G.P., T.W. Bruulsema, R.J. Gehl, D. Kanter, D.L. Mauzerall, C.A. Rotz, and C.O. Willaims. 2013. Nitrogen–climate interactions in US agriculture. Biogeochemistry 114:41–70. doi:10.1007/s10533-012-9802-4

Sainju, U.M., and B.P. Singh. 2001. Tillage, cover crop, and kill-planting date effects on corn yield and soil nitrogen. Agron. J. 93:878–886. doi:10.2134/agronj2001.934878x

Sainju, U.M., B.P. Singh, and W.F. Whitehead. 2002. Long-term effects of tillage, cover crops, and nitrogen fertilization on organic carbon and nitrogen concentrations in sandy loam soils in Georgia, USA. Soil Tillage Res. 63:167–179. doi:10.1016/S0167-1987(01)00244-6

Salmerón, M., J. Cavero, D. Quílez, and R. Isla. 2010. Winter cover crops affect monoculture maize yield and nitrogen leaching under irrigated Mediterranean conditions. Agron. J. 102:1700–1709. doi:10.2134/agronj2010.0180

Sharpley, A.N. 1985. The selective erosion of plant nutrients in runoff. Soil Sci. Soc. Am. J. 49:1527–1534. doi:10.2136/sssaj1985.03615995004900060039x

Sharpley, A.N., and S.J. Smith. 1991. Effect of cover crops on surface water quality. In: W.L. Hargrove, editor, Cover crops for clean water. Soil Water Conserv. Soc., Ankeny, IA. p. 1–91.

Singer, J.W. 2008. Corn Belt assessment of cover crop management and preferences. Agron. J. 100:1670–1672. doi:10.2134/agronj2008.0151

Singer, J.W., S.M. Nusser, and C.J. Alf. 2007. Are cover crops being used in the US Corn Belt? J. Soil Water Conserv. 62:353–358.

Smeltekop, H., D.E. Clay, and S.A. Clay. 2002. The impact of intercropping 'Sava' snail medic on corn production. Agron. J. 94:917–924. doi:10.2134/agronj2002.9170

Stahl, L. 2016. Managing risk when using herbicides and cover crops. Univ. of Minnesota Extension, St. Paul MN.

Strock, J.S., P.M. Porter, and M.P. Russelle. 2004. Cover cropping to reduce nitrate loss through subsurface drainage in the Northern U.S. Corn Belt. J. Environ. Qual. 33:1010–1016. doi:10.2134/jeq2004.1010

Sullivan, D.M., and N.D. Andrews. 2012. Estimating plant-available nitrogen release from cover crop. Pacific Northwest Extension Publ. PWW 636. https://catalog.extension.oregonstate.edu/sites/catalog/files/project/pdf/pnw636.pdf (accessed 7 Oct. 2016).

Uusitalo, R., E. Turtola, T. Kauppila, and T. Lilja. 2001. Particulate phosphorus and sediment in surface runoff and drainflow from clayey soils. J. Environ. Qual. 30:589–595. doi:10.2134/jeq2001.302589x

Wagger, M.G. 1989. Cover crop management and nitrogen rate in relation to growth and yield of no-till corn. Agron. J. 81:533–538. doi:10.2134/agronj1989.00021962008100030028x

Wagger, M.G., and D.B. Mengel. 1988. The role of nonleguminous cover crops in the efficient use of water and nitrogen. In: W.L. Hargrove, B.G. Ellis, T. Cavalier, R.C. Johnson, and R.J. Reginato, editors, Cropping strategies for efficient use of water and nitrogen. ASA Spec. Publ. 51. ASA, CSSA, and SSSA, Madison, WI. p. 115–127. doi:10.2134/asaspecpub51.c7

Wortman, S.E., C.A. Francis, M.L. Bernards, R.A. Drijber, and J.L. Lindquist. 2012. Optimizing cover crop benefits with diverse mixtures and an alternative termination method. Agron. J. 104:1425–1435. doi:10.2134/agronj2012.0185

- Estimation of Nitrous Oxide Fluxes from Agricultural Systems
- Factors Affecting Nitrous Oxide Emissions
- Role of Management in Reducing Nitrous Oxide Emissions

Soil and Nitrogen Management to Reduce Nitrous Oxide Emissions

Jerry L. Hatfield

Nitrous oxide (N_2O) emissions from agricultural soils represent a complex interaction between the inputs of nitrogen (N) into the soil and the soil environment. Mitigating these emissions will have a positive impact on greenhouse gases. Agriculture is the primary source of N_2O emissions and farmers must develop solutions to reduce these losses to the atmosphere. Inputs of nitrogen vary by form, rate, timing, and placement, and N_2O emissions are directly related to N inputs; however, the actual rate is governed by soil water content, soil temperature, soil carbon content, biological activity, and soil N content. The relationship with soil water shows an exponential increase when the water-filled pore space (WFPS) increases above 60 to 70% because of denitrification process, but there are situations in which N_2O can be released through nitrification with drier soils. Management practices which improve soil water balance by decreasing water content at times in which there is high soil N content, especially NO_3^--N, will reduce emissions along with improving oxygen exchange between the soil and the atmosphere. Practices using cover or catch crops to reduce both the soil N and the content show a positive impact, and altering forms of N by applying stabilizers or coatings can reduce emissions when coupled with water management. A wide range of practices, for example, manure and catch crops, show a positive yield-scaled benefit because these practices improve crop production. There are many opportunities surrounding N management with a potentially positive impact on reducing N_2O emissions, but these solutions may be regionally specific.

Nitrous oxide is a greenhouse gas directly related to nitrogen dynamics in the soil, especially the denitrification component in the nitrogen cycle. It is estimated that the lifetime of N_2O in the atmosphere is 114 yr with a greenhouse gas potential of 310 times CO_2, thereby constituting an extremely important component in greenhouse gas emissions. This has been well documented, and the most recent greenhouse gas inventory shows N_2O to be 6% of the total greenhouse gas emissions in the United States, with agriculture credited with 75% of the

---

Abbreviations: EF, emission factor; EEF, enhanced efficiency fertilizers; SOC, soil organic carbon; WFPS, water-filled pore space.

Jerry L. Hatfield, National Laboratory for Agriculture and the Environment, 2110 University Blvd., Ames, IA 50011 (jerry.hatfield@ars.usda.gov).

doi:10.2134/soilfertility.2014.0012

contributions within the United States (USEPA, 2014). Nitrous oxide is produced in agricultural soils by microbial transformations of N (fertilizers and manure). The primary factor affecting N_2O emissions is a combination of microbial and chemical processes as described in the following transformation:

$$NH_4^+ \rightarrow NH_2OH \rightarrow NO_2^- \rightarrow NO_3^- \qquad [1]$$

The two major processes producing N_2O are nitrification and denitrification. In nitrification there is the transformation of NH_4^+ to NO_3^- with N_2O emitted as byproduct. In denitrification there is a sequential reduction of NO_3^- to NO_2^- to N_2O and finally to N_2. Production of N_2O either by nitrification or denitrification reveals that these emissions can occur under a wide range of environmental conditions in field environments. When oxygen is limiting, ammonium oxidizers use NO_2^- as an electron acceptor and produce N_2O as described by Granli and Bøckman (1994). The dynamics of N_2O emission from the soil represents the result of nitrogen inputs, and the environmental conditions in the soil and this interaction has led to different methods developed to estimate emissions. De Klein et al. (2006) described the methodology used by IPCC to estimate N_2O emissions to provide a common basis of accounting greenhouse gas emissions, and these methodologies provide a framework to evaluate the role of N management in reducing N_2O emissions. Mosier et al. (1998) stated that an accurate global N_2O budget needs to account for all N inputs into agricultural systems, including synthetic fertilizer, animal waste, increased biological N-fixation, cultivation of mineral and organic soils through enhanced organic matter mineralization, and mineralization of crop residue returned to the field. Emissions of N_2O can occur directly from agricultural fields, animal confinements, or pastoral systems, and if N is transported sources into ground and surface waters and then released as N_2O, this emission would be considered an indirect emissions. The focus of this chapter is to evaluate crop and nitrogen management strategies that can potentially be implemented to reduce N_2O emissions from agricultural systems.

Estimation of Nitrous Oxide Fluxes from Agricultural Systems

There are a range of methods potentially employed to estimate direct N_2O emissions from soil. De Klein et al. (2006) classified direct emission sources as synthetic N fertilizers (F_{SN}), organic N applied as fertilizer (e.g., animal manure, compost, and sewage sludge) (F_{ON}), urine and manure N deposited directly onto pasture, range and paddock by grazing animals (F_{PRP}), N in crop residues (aboveground and belowground), including from N-fixing crops and from forages during pasture renewal (F_{CR}), N mineralization associated with loss of soil organic matter resulting from change of land use or management of mineral soils (F_{SOM}), and drainage and management of organic soils (i.e., Histosols) (F_{OS}). The direct factors are incorporated into the following equation:

$$N_2O_{Direct}-N = \\ N_2O-N_{N\ inputs} + N_2O-N_{OS} + N_2O-N_{PRP} \qquad [2]$$

where $N_2O_{Direct}-N$ are the total annual direct N_2O-N emissions produced from managed soils, kg $N_2O-N\ yr^{-1}$; N_2O-N_N inputs are annual direct N_2O-N emissions from N inputs to managed soils, kg $N_2O-N\ yr^{-1}$; N_2O-N_{OS} are annual direct N_2O-N emissions from managed organic soils, kg $N_2O-N\ yr^{-1}$; and N_2O-N_{PRP} are annual direct N_2O-N emissions from urine and manure inputs to grazed soils, kg $N_2O-N\ yr^{-1}$ (De Klein et al., 2006). In agricultural systems, Eq. [2] can be simplified to include only the N_2O-N_N inputs terms expressed as the inputs of $F_{SN} + F_{ON} + F_{CR} + F_{SOM}$ multiplied by an emission factor (EF_1). The inputs are expressed as kg N yr^{-1}, and the emission factor (EF) is assumed to be 0.1 with an uncertainty range of 0.003 to 0.03 (Bouwman et al., 2002a, 2002b). Lesschen et al. (2011) showed there was variation in the EF values depending on management, crop, and environment. Kim et al. (2013), after conducting a meta-analysis of research on EF values, concluded that the EF values were not linear and proposed a more curvilinear approach, as shown in Fig. 1. Their analysis revealed EF values were not consistent across rates, and when N rates were adequate to meet agronomic requirements there was a linear relationship between rate and emissions; however, rates above the agronomic optimum began to show a rapid increase due to excessive N in the soil (Fig. 1). Accounting for direct N_2O emissions will depend on a combination of local factors and comparison of management

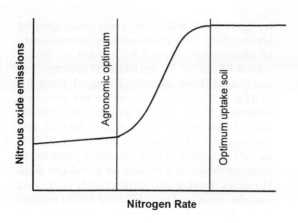

Fig. 1. Conceptual model of nitrous oxide emission response relative to nitrogen rate (see Kim et al., 2013).

practices requiring direct measurements of the emissions rather than utilizing a simple model.

Another approach developed for N_2O emissions for global-scale estimates is referred to as the Tier 2 method, expressed as

$$N_2O_{direct} - N =$$
$$\sum_i (F_{SN} + F_{ON})_i\, EF_{1i} + (F_{CR} + F_{SOM})EF_1 + \quad [3]$$
$$NO_2 - N_{OS} + NO_2 - N_{PRP}$$

where i represents different conditions in which the inputs are applied and are fully described by De Klein et al. (2006). These accounting methods for direct emissions provide a methodology for estimating the N_2O emissions from agricultural systems by including all of the potential N sources. These formulas provide insights into the description of the accounting tools but do not provide local-scale assessments of the variation in the EF which can be used to reduce N_2O release into the environment. The more detailed approach is described as Tier 3, methods which rely on models to estimate N_2O emissions or direct measurements from a specific management practice. Mu et al. (2009) developed an empirical model based on a compilation of observations from a range of experiments on the N_2O and CO_2 emission data concomitantly measured from agricultural upland fields around the world. This model is given as

$$\sum N_2O_e = a\, \exp\left[b * \left(\frac{E_{CO_2}}{S_{cn}} + F_n \right) \right] \quad [4]$$

where $\sum N_2O_e$ is the cumulative emissions of N_2O over a surface, a and b are empirical coefficients, E_{CO_2} is the emission of CO_2 from the soil,

S_{cn} is the soil carbon/nitrogen ratio, and F_n the input of N fertilizer. Mu et al. (2009) showed there was a difference between fertilized and unfertilized soils. They proposed that this type of model could be used to estimate N_2O emissions from soils as part of global-scale assessments from agricultural soils. Mu et al. (2009) found N_2O emission from soils is quite small and extremely variable in space and time which makes it difficult to directly estimate. In their model, they incorporated soil CO_2 emission which is related to soil temperature and moisture and is relatively easy to predict (Howard and Howard, 1993; Lloyd and Taylor, 1994). There have been other models developed, such as those proposed by Grant and Pattey (2003), Metivier et al. (2009), and Uzoma et al. (2015) in which these models simulate soil water dynamics and couple those dynamics with nitrogen responses under different management systems. The application of the model to allow for a more thorough examination of the spatial and temporal dynamics of the processes affecting N_2O emissions is valuable, as is illustrated by Metivier et al. (2009) in which they used the *ecosys* model to predict the N_2O emissions from a canola (*Brassica napus* L.) crop and found fertilizer application, precipitation, and temperature the main factors influencing N_2O emissions. Using this model they were able to estimate the temporal changes in N_2O emission similar to measured values and evaluate the potential impacts of different management practices on the emissions from this cropping system. However, measurements provide the direct evidence of the impact of changing nitrogen management systems on N_2O emissions, which is the primary source of information for this chapter.

Factors Affecting Nitrous Oxide Emissions

Lesschen et al. (2011) observed that environment (*E*), crop (*C*), or management (*M*) interact to determine EF values for N_2O emissions. Using either the Tier 1 or Tier 2 approaches we can change the rate (F_{SN}) or the EF as a $f(E,C,M)$ to achieve a reduction in N_2O emissions. The most simple solution to reduce N_2O would be to reduce inputs into the system; however, that may not be the most effective or optimal approach to achieve a meaningful reduction relative to expected crop production goals. Dobbie et al. (1999) observed that emissions of N_2O were affected by changes in the soil water status and temperature (representing *E*) and nitrogen application (*M*). These factors need to be evaluated before we can determine how to manage nitrogen to achieve effective solutions to reduce N_2O emissions. In the next sections, the factors will be dissected.

Soil Physical Environment

One of the most critical soil variables affecting N_2O emissions is soil water. Dobbie et al. (1999) found N_2O emissions were highest when the WFPS was greater than 70%, suggesting that denitrification was the primary factor. In a later study, Dobbie and Smith (2003) compared N_2O emissions from fertilized grassland with arable land across multiple sites in Great Britain for 3 yr and found emissions varied within a year and across sites and proposed an exponential relationship between WFPS and N_2O emission (Fig. 2). They found the major factors affecting N_2O emissions were soil water-filled pore space, temperature, and soil $NO_3–N$ content and observed large variations in EF values

for grassland with the same management across years. They recommended several years of observations would be needed to obtain robust EF values for agricultural systems. Values for EF from grassland ranged from 0.4 to 12.8 g N_2O-N ha^{-1} d^{-1}, while the variation observed for the cereal production systems ranged between 0.4–1.6 g N_2O-N ha^{-1} d^{-1} and for vegetables between 1.6 and 12.8 g N_2O-N ha^{-1} d^{-1}, suggesting that vegetable production systems behaved similarly to grassland sites. However, Dobbie and Smith (2003) suggested that the variation they observed would require separate EF values for each management system. In the grassland sites, the emissions were related to the amount of rainfall 1 wk before fertilizer application and the 3 wk following application. Emission factor values exponentially increased when the soil water-filled pore space was greater than 60% and soil $NO_3–N$ was not limiting (Fig. 2). The relationship they observed with prior rainfall demonstrates the need to carefully consider the environmental conditions when evaluating EF values for different management practices. The effect of soil water is not consistent because Mahmood et al. (1998) earlier had observed no significant relationship between soil water content and N_2O emissions. Choudhary et al. (2001), after comparing N_2O emissions from maize and permanent pasture, concluded soil water was the primary factor controlling emissions, and when the soil water content was less than 30% there was low emissions. This is further complicated by the spatial variation which exists within row crop systems in the N_2O emissions, as observed by Cai et al. (2012). In a series of experiments on maize with measurements made in the interrow and interrow plus row they found cumulative N_2O emissions varied from 0.84 to 1.22 kg N ha^{-1} from the interrow

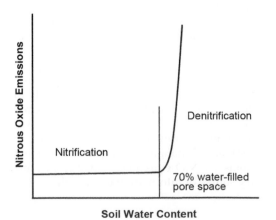

Fig. 2. Conceptual model of nitrous oxide response to soil water content.

plus row and decreased to 0.30 to 0.49 kg N ha[-1] from the interrow only and was further magnified by the N source. Manure application showed a larger spatial difference than inorganic N while there was no spatial difference from the control treatment. They suggested that measurements from only the interrow would underestimate the seasonal N_2O emissions and care needs to be taken when interpreting experimental results (Cai et al., 2012). The spatial variation of soil water between the row and interrow would further enhance the potential differences between these areas under a row crop.

Bateman and Baggs (2005) used a combination of stable isotope and acetylene (0.01% ,v/v) inhibition techniques to quantify N_2O production during denitrification, autotrophic nitrification, and heterotrophic nitrification in a fertilized (200 kg N ha[-1]) silt loam soil at contrasting (20–70%) WFPS. They observed that all of the N_2O emitted at 70% WFPS (31.6 mg N_2O-N m[-2] over 24 d or 0.16% of N applied) was produced via denitrification. When the WFPS was reduced to 35 to 60%, nitrification became the main process producing N_2O, and when WFPS was further reduced to 20%, denitrification became the dominant process affecting N_2O-N emissions and related to aerobic denitrification (Bateman and Baggs, 2005). They concluded that the combined stable isotope and acetylene approach to quantify the effect of WFPS on the processes affecting N_2O emission from agricultural soils proved to be a valuable technique. The process of denitrification is facilitated by a wide range of heterotrophic bacteria and fungi capable of reducing NO_3^- and NO_2^- to N_2O or N_2 under anaerobic conditions (Knowles, 1982; Shoun et al., 1992). This process is often considered to be the main N_2O producing process in soils because many studies have shown N_2O emissions after N application increase exponentially above 60% WFPS (Dobbie et al., 1999; Dobbie and Smith, 2003; Bateman and Baggs, 2005). As Bateman and Baggs (2005) observed, aerobic denitrification may provide substantial N_2O emissions in soils where O_2 is not limiting, which was observed by Patureau et al. (2000) because many bacteria isolated from soils and sediment (including *Pseudomonas*, *Aeromonas* and *Moraxella* genera) are capable of nitrate respiration in the presence of O_2 (Carter et al., 1995; Patureau et al., 2000). Emissions of N_2O from soil occurs over a large range of WFPS values, and the interactions with N management and soil water are key to understanding the temporal dynamics of N_2O emissions.

Soil Management Practices to Reduce Nitrous Oxide Emissions

There are several possibilities for management practices which may be effective in reducing N_2O emissions to the atmosphere. Based on either a Tier 1 or 2 approach, or any of the models, the primary factors directly related to management will be management of soil water, availability of NO_3-N in the soil, or the dynamics of the microbial activity. Interactions among nutrient management, cropping system, soil water, and the meteorological conditions all affect N_2O emissions with variations evident in the magnitude of the emissions among locations, experiments, and management systems.

Soil Water Management

Soil water dynamics represent one of the most complex factors affecting N_2O emissions and controlling soil water may offer one method to reduce N_2O emissions from agricultural systems. Recent studies by Maharjan et al. (2014) in Minnesota showed averages across all N sources (coated and uncoated fertilizers and single vs. split applications) no significant effect on N_2O emissions, which they attributed to very small differences in WFPS among irrigation treatments. In an earlier study using irrigation, Horváth et al. (2010) observed that N_2O emissions were 70% greater in irrigated plots compared to nonirrigated because the WFPS was nearly twice in the dry year but in a wetter year, there was no difference in WFPS or N_2O emissions. Welzmiller et al. (2008) observed that N_2O emissions increased after irrigation and showed that large increases began to occur when the WFPS was above 70%, and this factor was the most dominant in determining N gas emissions of both N_2 and N_2O from sorghum (*Sorghum biocolor* L. Moench.) experiments in Arizona. Scheer et al. (2008), after measuring N_2O emissions from cotton (*Gossypium hirsutum* L.) with a combination of irrigation and N management practices, observed that emissions occurred as pulses related to irrigation applications. These pulses produced very large emission rates, as high as 3 mg N_2O-N m[-2] h[-1] following fertilizer application and irrigation applications, and accounted for nearly 95% of the total emissions. These studies produced EF values of 0.0148 for this area near the Aral Sea Basin (Scheer et al., 2008), typical of other observed values in semiarid areas.

Cui et al. (2012) following a study on a silt loam calcareous soil in the North China Plain using irrigated maize–wheat (*Zea mays* L. and *Triticum aestivum* L.) cropping system showed that N_2O emissions after fertilization and irrigation and/or rainfall contributed to 73 and 88% of the annual N_2O and NO emissions, respectively, and were affected by soil moisture and mineral N contents similar to previous results. The total annual emissions from fields fertilized at the conventional rate (600 kg N ha^{-1} yr^{-1}) were 4.0 ± 0.2 and 3.0 ± 0.2 kg N ha^{-1} yr^{-1} for N_2O and NO, respectively, while those from unfertilized fields were much lower (0.5 ± 0.02 kg N ha^{-1} yr^{-1} and 0.4 ± 0.05 kg N ha^{-1} yr^{-1}, respectively). Cui et al. (2012) estimated that the EF values for N_2O and NO for urea fertilizer nitrogen were $0.59 \pm 0.04\%$ and $0.44 \pm 0.04\%$, respectively, which was near the lower range of values observed by Dobbie and Smith (2003). Cui et al. (2012) stated that nitrification dominated the emission processes for both N_2O and NO following N fertilization because there was insufficient available carbon for microbial denitrification to drive N_2O emission, and concluded that soil management practices directed toward increasing the carbon storage sink in calcareous soils may lead to an increase in N_2O emissions.

To further illustrate the temporal dynamics of N_2O emissions, there have been observations of emissions under different irrigation management systems by Sánchez-Martín et al. (2008a, 2008b). In their study on melons (*Cucumis melo* L. "Sancho"), they compared furrow vs. drip irrigation by using a comparison between a single rate of ammonium sulfate (175 kg N ha^{-1}) compared with no fertilizer. The addition of the fertilizer increased the N_2O emissions for both irrigation management systems; however, the emissions from the drip irrigation system were reduced by 70% compared with the flood irrigation system. They attributed this effect to the lower amount of water applied and differences in the wetting pattern of the soil. In the sequence of measurements following the fertilizer application, the dry area and wet area of the drip irrigation system showed the same emission rate (0.01 g N_2O-N m^{-2} d^{-1}); however, there were notable pulses of N_2O emitted from the wet areas throughout the season, whereas the pulse from the dry area was only detectable immediately following application. An interesting observation in this experiment was that the dry areas of the drip-irrigated plots emitted a similar amount of N_2O compared with the wet areas (0.45 kg N_2O-N ha^{-1}) in the no-fertilizer plot but showed greater total emissions in the fertilized plot area (0.92 kg N_2O-N ha^{-1} for dry vs. 0.70 kg N_2O-N ha^{-1} for wet). They attributed the pulses of N_2O to the frequent increases in the soil wetting volume after the addition of water in the drip irrigation system and further suggested that the primary factor affecting these emissions was nitrification under the drip system but denitrification under the flood irrigation system. They concluded that drip irrigation would be an effective management tool to reduce irrigation water requirements and mitigate N_2O emissions. Sander et al. (2014), in a study on rice (*Oryza sativa* L.) in Philippines, evaluated both the within-season N_2O emissions and during the fallow period and observed very little N_2O emission during the growing season when the rice was flooded. During the off-season, N_2O emissions were affected by residue management, with the removal of the residue creating a larger emission rate. The interaction between soil water and residue further increases the need to quantify the sources of carbon and the interactions with microbial activity in the soil.

Aguilera et al. (2013) conducted a meta-analysis for experiments conducted across Mediterranean environments and found a large range in EF values. Synthetic plus organic fertilizers had EF values of 3.04 while synthetic alone was 1.71, organic solids were 0.91, and organic liquid fertilizers were 1.75. Water had a much larger effect with rain-fed conditions showing EF values of 0.09 while high water regimes and low water application averaged 2.02 and 1.65, respectively. Fertilizer rate behaved as expected with rates from 0 to 75 kg ha^{-1}, exhibiting an EF value of 0.06, 75–150 kg ha^{-1} at 0.39, and 150–225 kg ha^{-1} at 2.31. There were differences in cropping systems; however, these are governed by the amount of fertilizer applied and the water regime with maize and showing an EF value of 2.74 and 0.10, respectively. They observed across these studies that indirect emissions have not been fully accounted for, but when organic fertilizers are applied at similar N rates to synthetic fertilizers, they generally make smaller contributions to the leached NO_3–N pool and would reduce the net N_2O emissions. They summarized that the "most promising practices for reducing N_2O through organic fertilization include: (i) minimizing water applications; (ii) minimizing bare soil; (iii) improving waste management; and (iv) tightening N cycling through N immobilization. The mitigation potential may be limited by: (i) residual effect; (ii) the long-term effects of fertilizers on soil organic matter; (iii) lower yield-scaled performance; and (iv) total N availability from organic sources." Knowledge gaps identified in their review included: "(i) insufficient sampling periods; (ii) high background emissions; (iii) the need to provide N_2O EF and yield-scaled EF; (iv) the need for more research on specific cropping systems; and (v) the need for full GHG balances." (Aguilera et al., 2013). This review provides a stage for the challenges in evaluating practices to impact N_2O emissions. The effectiveness of these practices vary among research study for many different factors, and the potential of impacting N_2O emissions needs be evaluated across cropping systems, climates, and management systems.

In comparing different forms of fertilizer and tillage systems, Halvorson et al. (2014) showed that N_2O emissions that were a function of N applied with N_2O emissions (g N ha^{-1}) were almost four times larger in the extremely wet year (slope of the regression equals 15.44 × N rate) compared with the normal years (slope of the regression equals 4.30 × N rate). These observations suggest that in the wet year, there was a greater likelihood of the soil exceeding the critical WFPS threshold than in the more typical years for this Colorado location. This further reinforces the concept that multiple years of observations may be needed to bracket the expected variation in soil water dynamics related to N_2O emissions. Soil water has been found to be the dominant factor affecting the temporal N_2O emissions from soil.

Soil Nutrient Management

One of the primary factors affecting N_2O emissions is the addition of N sources to the soil and its effect on the NO_3–N availability in the soil profile. Nutrient management can be separated into rate, form, timing, or placement effects; however, these factors determine the availability of NO_3–N in the soil profile. Evaluations conducted on different aspects of nutrient management provide insights into the role of nutrient management on N_2O emissions. Comparison studies on organic vs. synthetic fertilizers and enhanced efficiency fertilizers vs. nonenhanced forms of the same material have shown different effects on N_2O emissions.

Effect of Rate on Nitrous Oxide Emissions

Changing N rate is often considered to be the most direct route to reducing emissions based on Eq. [2, 3, or 4]. Chirinda et al. (2010) compared N_2O emissions from soil under winter wheat within three organic and one conventional cropping system that differed in amount of N applied combined with the type of fertilizer, presence of catch crops, and proportion of N_2–fixing crops on sandy loam soils under different climatic conditions in Denmark. The conventional rotation received 165 to 170 kg N ha^{-1} in the form of NH_4NO_3, while the organic rotations received 100 to 110 kg N ha^{-1} as pig manure. Mean daily N_2O emissions across the year ranged from 172 to 438 mg N m^{-2} d^{-1} at the wetter site and from 173 to 250 mg N m^{-2} d^{-1} at the drier site. There was a significant inter-annual variation in N_2O emissions; however, there was no significant difference between the organic and conventional systems even though there was less N input in organic rotations. The annual emissions were equivalent to 0.5–0.8% of the N applied as either manure or mineral fertilizer. In their study, they observed that the high N_2O emissions were coincident with high soil NO_3–N levels or high soil water content (high WFPS). Dittert et al. (2005) compared mineral vs. organic fertilizers in the early spring for a grassland system in Germany and found that N_2O emissions were affected by the combination of manure and commercial fertilizer with fresh cattle manure combined with calcium ammonium nitrate mineral fertilizer increasing N_2O flux during the first 4 d after application from 10 to 300 mg N_2O–N m^{-2} h^{-1}. They used ^{15}N labeled manure or fertilizer to evaluate the emitted N_2O to demonstrate that manure C created an increase in N_2O–N emission from mineral fertilizer and that N_2O fluxes from either N source alone showed less of an increase than the combination. Their research points out the observation that attention needs to be directed toward the N pools and the interactions among N forms in the soil for different N materials (Dittert et al., 2005). In a companion study, Lampe et al. (2006) utilized ^{15}N-labeled materials to evaluate N_2O emissions from permanent grassland in Germany to compare the commercial rates (0 and 100 kg N ha^{-1} yr^{-1}) with two manure rates (0 and 74 kg ha^{-1} yr^{-1}) across two harvests and two grazing cycles. Mean cumulative N_2O-N loss was 3.0 kg ha^{-1} yr^{-1}, and N_2O emissions varied from 0.03 to 0.19% of the N applied. The labeled ^{15}N materials showed that mineral N emitted more N_2O than the manure treated plots because of the greater soil N pool in the fertilizer plots. In this study, there were no differences among treatments for the cumulative N_2O losses which they attributed to the large variation in emissions caused by the manure deposited from the grazing animals and the N_2 fixation by the white clover (*Trifolium repens* L.) in the 0 N rate (Lampe et al., 2006). In their observations, N_2O emission rates of up to 147 mg N_2O-N m^{-2} h^{-1} were measured during freeze–thaw cycles in winter (December–February), which was 26% of the annual N_2O flux.

Annual N_2O emissions revealed that N fertilizer application and grazing create short-term increases of N_2O emissions while the major portion of the annual N_2O emissions can be attributed to the soil N pool (Lampe et al., 2006). The significant portion of N_2O emissions occurring during the freeze–thaw cycles suggest that it would be necessary to measure these portions of the year to obtain an accurate measure of annual N_2O emissions. A similar conclusion was reached by Singurindy et al. (2009) after observing that N_2O emissions during soil thaws in winter and spring in New York were a significant portion of the total annual emissions from agricultural lands. Their results showed that tillage reduced N_2O emissions in nonmanured soils by 20 to 30% in the 35 to 50 d following the tillage event, which they attributed to improved aeration caused by reduced bulk densities and pore space saturation. In this study, N_2O emissions of 200 μg N m^{-2} h^{-1} were found at soil temperatures greater than 5°C and at WFPS between 40 and 70%. Emissions of N_2O from manure-amended soils were not limited to thawing events, and emissions were observed to begin at soil temperatures below 0°C and continued after the soil was frozen. Their observations revealed that previous tillage operations before the manure application had a significant impact on N_2O emissions during freeze–thaw cycles following manure application, with the larger emissions coming from fields tilled earlier in the fall. As in other studies, there was an increase in N_2O emissions, with increasing WFPS with 60 to 70% showing the largest effect on N_2O emissions. There is a complex set of interactions between tillage and fertilizer management which highlights the effects of N pools and aeration of the soil, as demonstrated by several investigators (Dobbie and Smith, 2003; Dittert et al., 2005; Lampe et al., 2006; Singurindy et al., 2009). De Antoni Migliorati et al. (2014) found that four different rates of urea applied to maize had N_2O emissions in proportion to the N applied. Urea was applied at 40 kg N ha^{-1}, 120 kg N ha^{-1}, 240 kg N ha^{-1}, and 240 kg N ha^{-1} plus DMPP (3,4-demethylpyrazole phosphate) as an inhibitor and produced annual N_2O emission values of 0.49, 0.84, 2.02, and 0.74 kg N_2O-N ha^{-1} yr^{-1}. This verifies the assumption that N rate directly affects N_2O emissions, and

one caution these authors raised was that the adjustment of N rates based on residual soil N values requires an accurate assessment of soil N to avoid yield reduction due to inadequate N for the crop.

Forms of Nitrogen on Modifying Nitrous Oxide Emissions

Comparisons of different forms of N have shown mixed results on N_2O emissions and related to the availability of NO_3 in the soil profile and the interaction with WFPS. Hayakawa et al. (2009) evaluated the influence of pelleted poultry manure on N_2O and NO fluxes from an Andisol soil in Japan by comparing poultry manure, pelleted poultry manure, and chemical fertilizer (NH_4–N [$(NH_4)_2SO_4$]) at 120 kg N ha^{-1} in each cultivation period to rape (*Brassica rapa* var. *peruviridis*). The highest N_2O flux was observed from the pelleted poultry manure after a rainfall following fertilization, and the cumulative emission rate (2.72 ± 0.22 kg N ha^{-1} y^{-1}) was 3.9 times greater than from the poultry manure and 7.1 times compared with commercial fertilizer treatments. The pelleted poultry manure produced an EF value of 0.013 in this study. Emission rates of NO were highest from the commercial fertilizer and driven by nitrification. Cumulative N_2O emission rates from all N fertilizer treatments were higher during the wetter period compared with the drier cultivation period. In contrast, NO emission rates were higher in the drier than in the wetter cultivation period. In their incubation study on N sources, WFPS values above 50% produced the highest N_2O and CO_2 fluxes and the lowest soil NO_3^- content (Hayakawa et al., 2009). They attributed the differences among fertilizer materials to the anaerobic environment inside the pellets, caused by rainfall and heterotrophic microbial activities, which in turn promoted denitrification, leading to an increase in N_2O emissions. They suggested that one strategy to reduce emissions would be to manage the timing of N application to avoid wet soil conditions (Hayakawa et al., 2009). Martins et al. (2015) compared urea, urea + zeolite, calcium nitrate, and ammonium sulfate at 120 kg N ha^{-1} as topdress treatments and found no significant differences for the EF values under a maize crop. Their estimated EF values were 0.2 compared with the Tier 1 value of 1%, and they and suggested that use of the IPPC approach would greatly overestimate N_2O emissions.

After measuring N_2O emissions in turfgrass grown near Manhattan, KS, with different forms of N fertilizer (urea and ammonium sulfate), Bremer (2006) concluded that N fertilization increased N_2O emissions by as much as 15 times within 3 d. Precipitation was a dominant factor affecting N_2O emissions because there was a variation in emissions after each application, depending on the amount of precipitation occurring within 2 to 3 d after application. In this study, the N_2O fluxes were $-22\ \mu g\ N_2O\text{-}N\ m^{-2}\ h^{-1}$ during the winter to $407\ \mu g\ N_2O\text{-}N\ m^{-2}\ h^{-1}$ after fall fertilization. The temporal dynamics of N_2O emissions showed fluxes increased for 2 to 3 wk after application and then decreased to background concentrations. Over the year, the cumulative annual emissions of $N_2O\text{-}N$ were 1.65 kg ha^{-1} for the high rate of urea (250 kg ha^{-1} or 0.066 EF), 1.60 kg ha^{-1} in ammonium sulfate (250 kg ha^{-1} or 0.064 EF), and 1.01 kg ha^{-1} in a low rate of urea (50 kg ha^{-1} or 0.0202 EF). Bremer (2006) concluded that the high rate of urea increased annual N_2O emissions in turfgrass by 63% compared with the low rate of urea, with no difference between urea and ammonium sulfate at the 250 kg ha^{-1} rate. In turfgrass, the form of N does not appear to be a factor affecting N_2O emission, with the more dominant factor being timing and amount of precipitation relative to N application.

Placement of Nitrogen in Affecting Nitrous Oxide Emissions

Placement of N into the soil as either broadcast incorporated or banded compared with a surface application has the potential to affect the N_2O emissions primarily through the interaction with the soil water and soil $NO_3\text{-}N$ pools. Placement of N into bands in the soil has the potential to increase N_2O emissions because of the greater $NO_3\text{-}N$ in the soil. Pfab et al. (2012) observed no effect of broadcast vs. banded placement, with the larger difference among treatments being the use of DMPP as an inhibitor on ammonium sulfate. This is in contrast to the findings of Smith et al. (2012) in comparing urea-ammonium nitrate, ammonium nitrate, and poultry litter at rates of 170 kg N ha^{-1} for maize production as either broadcast or banded, where banding produced the largest losses of CO_2 and N_2O. Poultry litter had the highest emissions compared with the inorganic fertilizer. Maharjan and Venterea (2013) showed that the placement and form of N materials had an effect on nitrite (NO_2^-) and NO_3^-, and the

differences between banding and broadcast or stabilized or unstabilized urea could be related to how these practices affected these pools. Zhu-Barker et al. (2015) observed that N_2O emissions were a function of fertilizer form and application method based on their comparison of knife-injected anhydrous ammonia with urea or calcium nitrate applied as a topdress material to wheat. There was an increase in N_2O emissions with broadcast urea compared with calcium nitrate because of the effect of ammonia oxidation on N_2O generation. There is an interaction with the form or N supplied and the placement of these materials relative to the soil water dynamics of the cropping system.

Role of Enhanced Efficiency Fertilizers in Reducing Nitrous Oxide Emissions

One of the aspects of nutrient management has been the introduction of enhanced efficiency fertilizers (EEF) with the goal of reducing N_2O emissions (Halvorson and Del Grosso, 2012). These products are defined by the Association of American Plant Food Control Officials (AAPFCO, 2013, Official Publication 65, T-70) as "fertilizer products with characteristics that allow increased plant uptake and reduce the potential of nutrient losses to the environment (e.g., gaseous losses, leaching, or runoff) when compared to an appropriate reference product." Enhanced efficiency reference products are currently defined by AAPFCO as "soluble fertilizer products (before treatment by reaction, coating, encapsulation, addition of inhibitors, compaction, occlusion, or by other means) or the corresponding product used for comparison to substantiate enhanced efficiency claims."

There have been several studies conducted evaluating the effect of EEF materials on N_2O emissions with mixed results. Halvorson et al. (2010a, 2010b) showed there was a positive effect of EEF materials on reducing N_2O emissions under irrigated conditions in Colorado. However, in rain-fed conditions, Sistani et al. (2011), Venterea et al. (2011), and Parkin and Hatfield (2014) found that EEF materials did not reduce the emissions compared with untreated materials. Sistani et al. (2011) and Parkin and Hatfield (2014) showed an increase in N_2O emissions late in the growing season, which they attributed to a greater amount of $NO_3\text{-}N$ in the soil profile. Parkin and Hatfield (2014) found large differences among years in the cumulative emissions caused by the differences in precipitation related to WFPS with no differences in the fertilizers, and in 2009 environmentally sensitive nitrogen (ESN), a polymer coated urea material, showed a significantly higher cumulative N_2O emission compared with other N sources. Over the 3 yr of their study, the EF values ranged from 0.017 to 0.07. Parkin and Hatfield (2014) proposed that the differences between rain-fed and irrigated systems are due to the seasonal dynamics of the soil NO_3–N concentrations in the soil profile. This effect was described by Asgedom et al. (2013), in which soil NO_3–N was the determining factor in N_2O emissions. Evaluation of different N sources may not show a consistent effect because of the interrelationship between soil water and soil NO_3–N. This is an extension of the observations from other studies in which WFPS was the determining factor in N_2O emissions (e.g., Dobbie and Smith [2003]). However, use of the EEF materials do show a positive effect on grain yield for a number of crops, which increase the yield-scaled N_2O emissions; for example, Halvorson and Bartolo (2014) showed significant increase in maize yield with EEF materials and suggested that less N would be required to achieve the same grain yield. A similar result was observed by Li et al. (2012) for canola in Canada, in which the application of ESN showed a decrease in N_2O emissions with no change in grain yield, resulting in an improvement in yield-scaled N_2O emissions. In this study, there was one site which showed an increase in N_2O emissions caused by unfavorable growing conditions for the canola which increased the available N in the soil profile. Following a study on broccoli (*Brassica oleracea*) in Australia, Scheer et al. (2014) concluded that the inhibitor 3,4-dimethylpyrazole phosphate (DMPP) was effective in reducing N_2O emissions by as much as 75% compared with the standard practice during the growing season, but over the course of the entire year there was no effect because of the impact of the broccoli residue on the soil NO_3–N concentrations. Zhang et al. (2015a) compared urea, urea + dicyandiamide, urea + nitrapyrin, and area + biological nitrification inhibitor using a rate of 1112–969–969 kg N-P_2O_5–K_2O kg ha^{-1} yr^{-1} on seven sequential vegetable crops [Amaranth (*Amaranthus mangostanus* L.), Tung choy (*Ipomoea aquatic* Forsk. L.), baby bok choy (*Brassica rapa* Chinensis L.), coriander herb (*Coriandrum sativum* L.)] over a 2-yr period to evaluate the effect of nitrification inhibitors on N_2O emissions and crop yields. Each crop was irrigated during the growing season to ensure no water deficits affected production.

They found no difference between the urea and urea + dicyandiamide in either production or N_2O emissions; however, the urea + nitrapyrin and urea + biological nitrification inhibitor decreased the annual N_2O emissions and increased the nitrogen use efficiency of the vegetable production system. Emissions of N_2O were reduced from 32.1 kg N ha^{-1} yr^{-1} in the urea treatment to 26.8 and 26.3 in the urea + nitrapyrin and urea + biological nitrification inhibitor, respectively. Over the 2 yr of this study, N_2O emissions were correlated with temperature (0.54), WFPS (0.40), and NO_3–N (0.60), which were all significant (Zhang et al., 2015a). There are positive effects of nitrification inhibitors on reducing N_2O emissions caused by a reduction in the NO_3–N pool. This study linked both soil water and NO_3–N pools to show that these two factors contribute to reducing emissions. The effectiveness of EEF materials needs to be evaluated relative to the soil temperature and WFPS, along with the soil NO_3–N pools over the entire growing season, to determine the overall impact. The overall conclusion based on EEF materials is that the effect of these materials on N_2O emissions resides in the ability to reduce the NO_3–N availability in the soil profile and to decrease the availability in the early season when crop uptake is low. This is an area which needs more attention to determine how best to manage N using EEF materials and balance crop N requirements and N_2O emissions.

Interaction of Nutrient Management and Tillage

Interactions between tillage and N management were evaluated by Plaza-Bonilla et al. (2014) for a Mediterranean environment, comparing conventional tillage and no-till with different forms of N. They compared N_2O emissions from a long-term experiment with two tillage systems and three N fertilizer rates (0, 60, and 120 kg N ha^{-1}) and then compared these same tillage systems with three N rates (0, 75, and 150 kg N ha^{-1}) and two sources of N (ammonium sulfate/ammonium nitrate and pig manure). In this study, they reported the N_2O emissions as kilograms of CO_2 equivalents per kilogram of grain produced as a measure of a yield-scaled metric for emissions. Across all of the fertilizer treatments the conventional tillage systems emitted 0.362 in the long-term study and 0.104 kg CO_2 equivalent kg grain^{-1} in the short-term experiment, while the no-till system emitted 0.033 and 0.056 kg CO_2 equivalent kg grain^{-1}, respectively. In this study, N fertilization rates did not affect the average N_2O fluxes or the total N losses in the long-term experiment. However, in the short-term experiment, N_2O emissions increased with application rate for both mineral and organic fertilizers, and the application of pig manure increased grain production compared with the mineral N treatment, causing a reduction in yield-scaled N_2O emissions by 44% (Plaza-Bonilla et al., 2014). Comparison of management systems using the yield-scaled approach may be a more effective method of comparing management systems for their true environment impact. This analysis demonstrated an advantage of no-till in reducing N_2O emissions; however, there have been suggestions that N_2O emissions would be increased through enhanced soil water content and more available C in these tillage systems (Aulakh et al., 1984; Ball et al., 1999; Smith et al., 2001; Tan et al., 2009; Sheehy et al., 2013).

Kong et al. (2009) observed that the transition between tillage systems showed an increase in N stabilization under reduced tillage, which contributed to a reduction in N_2O emissions. However, after a meta-analysis of 239 direct comparisons between conventional tillage and no-till or reduced tillage, Van Kessel et al. (2013) observed that no-till or reduced tillage did not produce higher N_2O emissions compared with conventional tillage after 10 yr of no-till practice. There was an advantage from no-till and reduced tillage systems over the long-term, and it was even more noticeable in dry climates. Venterea et al. (2005) in Minnesota found no-till systems reduced N_2O emissions compared with conventional tillage and conservation tillage. They argued that the interactions between tillage and N fertilization management need to be understood to interpret the observations from different experiments and to make recommendations on soil management systems capable of reducing N_2O emissions. Ussiri et al. (2009) compared no-till with chisel plow and moldboard plow in Ohio and found N_2O emissions were 1.82 kg N ha^{-1} yr^{-1} for moldboard plow, 1.96 kg N ha^{-1} yr^{-1} for chisel plow, and 0.94 kg N ha^{-1} yr^{-1} for the no-till system. They attributed the reduced emissions in the no-till to enhanced gas exchange, even though this tillage system had a higher water content and greater bulk density. This same effect was noted by Petersen et al. (2008), in which they found that relative gas diffusivity provided a representation of the soil water effect. Tan et al. (2009) concluded there was a significant interaction among crop rotation, tillage system, and timing of the N

application and that higher N_2O emissions in their studies were related to the fraction of large pores and C availability in the upper soil profile. Sheehy et al. (2013) observed that cumulative emissions of N_2O varied from 2.4 to 8.3 in conventional tillage, 0.5 to 6.5 in reduced tillage, and 4.9 to 10.2 kg N_2O-N ha^{-1} in no-till cereal systems in northern Europe. Ball et al. (2014) concluded that N_2O fluxes were only different at the highest levels of residue incorporation or when the rotation and crop interacted with rainfall patterns. They found that N_2O fluxes did not relate to rainfall differences among seasons; however, the patterns of rainfall increased the differences among tillage, organic residues, and phase of the crop rotation. They concluded that timing of the operations during dry periods would not be an effective N_2O mitigation strategy (Ball et al., 2014). Given the variation they observed in N_2O fluxes, they suggested that more intense measurement are necessary to overcome some of the variation observed among treatments.

Pelster et al. (2011) observed that in a maize–soybean system on a well-drained soil near Montreal, N fertilizer applications increased N_2O emissions while tillage and the tillage–fertilization interaction had no measurable effect. Their values of N_2O emissions from April through November ranged from 1 to 2.5 kg N_2O-N ha^{-1}, with losses of N fertilizer ranging between 0.9 to 1.3% with the differences being between the 2 yr of their study. Zhang et al. (2015b) evaluated different tillage systems in a wheat–rice production system in Yangtze River Valley of China only during the wheat portion of the rotation. They compared both N_2O and CH_4 emissions under no-till, reduced tillage, and conventional tillage systems. Over the 3 yr of their study, N_2O emissions for the no-till was 2.24 kg N_2O-N ha^{-1}, conventional tillage was 2.01 kg N_2O-N ha^{-1}, and only 1.73 kg N_2O-N ha^{-1} for the reduced tillage system which showed basically no effect of tillage systems on emissions (Zhang et al., 2015b). Choudhary et al. (2002) showed a similar result for a maize crop in New Zealand after comparing conventional tillage to no-tillage systems with 9.2 kg N_2O-N ha^{-1} yr^{-1} in conventional tillage and 12.0 kg N_2O-N ha^{-1} yr^{-1} compared to only 1.66 kg N_2O-N ha^{-1} yr^{-1} under permanent pasture. They found no difference between the conventional and no-tillage systems because of the large spatial variation within each tillage system (Choudhary et al., 2002). Liu et al. (2014) compared a combination of mulching practices obtained by the degree of tillage and N application rates for the effect on N_2O emissions and found that the effect of mulching on maize grain yield reduced the yield-scaled N_2O emissions. These results demonstrate that there are many paths to achieving a reduction in the N_2O emissions relative to improved crop performance, and the interactions among tillage, fertilizer, and weather determine the actual N_2O emissions.

Interaction of Cropping Systems with Tillage and Nutrient Management

There is an interacting effect among tillage systems, cropping systems, and N sources (Bayer et al., 2015). The results from an 18-yr old experiment in a subtropical Acrisol of southern Brazil evaluating the long-term effects of conventional tillage compared with no-tillage and two cropping systems consisting of black oat (*Avena strigose* Schreb.)–maize and a vetch (*Vicia sativa* L.)–maize system revealed that tillage systems did not affect N_2O emissions, but increased four times under vetch residues because of increase in soil NO_3–N, NH_4^+-N and dissolved organic C in the soil. Even though there was an increased N_2O emission in the no-till system with the vetch system, the increase in maize yield under this cropping system made this an effective strategy for reducing N_2O emissions because adding urea as an N source produced a threefold increase compared with the legume cover crop. They showed that the yield-scaled N_2O emissions were 67 g N Mg^{-1} with the legume residue and 152 g N Mg^{-1} with the urea fertilizer (Bayer et al., 2015), demonstrating that providing a portion of the maize N with winter legumes would mitigate N_2O emission in subtropical agriculture systems. Li et al. (2015) showed a similar effect from legume cover crops with red clover (*Trifolium pratense*), red clover with ryegrass (*Lolium perenne*), vetch (*Vicia villosa*), and fodder radish (*Raphanus sativus* L.), where the legume crops accumulated 63 kg N ha^{-1} in the aboveground material compared with only 36 kg N ha^{-1} in the nonlegumes which increased grain yield of the barley (*Hordeum vulgare* L.) crop by 30%. The annual N_2O emissions were similar across the crops, which demonstrates that legume crops create an environment in which there is enhanced effect of legume cover crops (Li et al., 2015; De Antoni Migliorati et al., 2015). Sauer et al. (2009) evaluated an interseeded rye (*Secale cereale* L.) cover crop in a bermudagrass (*Cynodon dactylon* L. Pers.) pasture coupled with poultry manure in Arkansas and observed N_2O emissions were positively correlated with soil

NO_3–N levels and soil water content. Additions of poultry manure caused an increase in N_2O emissions; however, this response was offset by the presence of the rye cover crop because the reduction in the soil NO_3–N levels and decrease in the WFPS. They concluded that the interplay between the addition of the poultry manure and the use of the rye cover crop could be combined to achieve environmental benefits and sustain productivity of this production system (Sauer et al., 2009).

Manalil and Flower (2014) observed N_2O emissions in Australia in both the cropping season and the fallow period and observed N_2O emissions recorded in this study were relatively low even after fertilization of the soil, with values of 0.18 g N_2O-N ha^{-1} h^{-1} (4.32 g N_2O-N ha^{-1} d^{-1}) similar to those observed in cropping regions of this area (Barton et al., 2008, 2010). In their study, maintaining the soil in a fallow condition did not cause an increase in nitrous oxide from planting to harvest the next year, compared with a continuous cropping system. They attributed the high rates of N_2O emission to a combination of the high temperatures increasing the rate of nitrification and the anaerobic environment enhanced by rainfall, as suggested by Harris et al. (2013). Similar to the synthesis of Aguilera et al. (2013), these results show that the combination of low temperatures during the wet growing seasons and dry summer conditions in the Mediterranean climate show that the peak emission of N_2O rarely exists during crop growing seasons where mineral fertilizer N is applied. The combination of soil water dynamics and fertilizer timing determine the dynamics of N_2O emissions throughout the year.

In onion (*Allium cepa* L.) production systems with different cover crops using either clover or a mixed herb system, which were incorporated into the 3 mo before the measurements commenced, the N_2O emissions were significantly greater from the clover cover crop compared with herb cover crop (van der Weerden et al., 2000). There were pulses of N_2O emissions following seedbed preparation and seeding, and soil water content was the controlling variable. They observed that adding N fertilizer to the clover system had both direct and indirect impacts on N_2O emissions by increasing the rate of nitrification, increasing the soil NO_3–N levels for potential denitrification, and changing the ratio of N_2O/N_2 by altering the soil pH (van der Weerden et al., 2000). During the season, the clover system increased onion yields, which improved the overall yield-scaled values;

however, the N_2O emissions during this part of the overall growing season were small compared with the period with cultivation, and there was no difference among the management systems in N_2O emissions. These results demonstrate the necessity to measure N_2O emissions through the complete year to compare production systems.

There may be differences among the varieties selected in evaluations of N_2O emissions. There are few studies with more than a single variety, and a study by Gogoi and Baruah (2012) in the northern plain of Assam, India, comparing wheat and rice varieties found differences among wheat and rice varieties in seasonal N_2O emissions. Emissions of N_2O emissions from different wheat varieties ranged from 12 to 291 μg N_2O-N m^{-2} h^{-1}, and seasonal N_2O emissions ranged from 312 to 385 mg N_2O-N m^{-2}. During the rice season, N_2O emissions ranged from 11 to 154 μg N_2O-N m^{-2} h^{-1}, with seasonal N_2O emission of 190 to 216 mg N_2O-N m^{-2}. The seasonal integrated flux of N_2O differed significantly among wheat and rice varieties. In the wheat growing season, N_2O emissions correlated with soil organic carbon (SOC), soil NO_3^--N, soil temperature, shoot dry weight, and root dry weight. The most significant variables governing N_2O emissions during the wheat study were soil temperature, followed by SOC and soil NO_3^--N. In the rice season, N_2O emissions were significantly correlated with SOC, soil NO_3^--N, soil temperature, leaf area, shoot dry weight, and root dry weight, with the primary variables affecting N_2O emissions being soil NO_3^--N, leaf area, and SOC. One of the primary differences among wheat and rice varieties with higher N_2O emissions showed a higher root biomass. They suggested that selection of varieties with reduced N_2O emissions would be a potential mitigation strategy.

In cropping systems, there are some general observations which demonstrate that N_2O emissions are affected by the interaction of tillage, nutrient management, cover crops, and the soil properties. These factors interact with the SOC, soil NO_3^--N concentrations, and soil water status. Cropping systems and their management affect all of these parameters throughout the year and contribute to the variation observed in N_2O emissions among management systems, within a year and among years. There are often short-term studies conducted on a limited temporal scale which may not fully capture the dynamics affecting N_2O emissions. To illustrate this point is the analysis presented by Decock (2014), in which she conducted a meta-analysis

of N_2O emissions from Midwestern US and Canada farming systems in which there 48 studies with cumulative N_2O emission, 18 comparing fertilizer induced emissions, 33 with crop yields, 7 which measured crop N export, and only 10 studies with year-round measurements. These studies are fairly sparse for drawing definitive conclusions on potential practices to reduce N_2O emissions. The temporal variation in N_2O emissions and the interaction with soil water dynamics suggest that to improve our understanding of management systems with the potential to reduce emissions will require measurements throughout the year.

Role of Management in Reducing Nitrous Oxide Emissions

Achieving a reduction in N_2O emissions is not as simple as simply changing the rate or form of N. Changing the amount of N being placed into the system certainly has a large impact (e.g., Bouwman et al., 2002a, 2002b; Dobbie et al., 1999; Dittert et al., 2005; Halvorson et al., 2010a, 2010b; Horváth et al., 2010; Plaza-Bonilla et al., 2014; Nayak et al., 2015; Snyder et al., 2009); however, the results from the studies on different cropping systems and soils show that there is an interaction of the N rate with the soil water availability (WFPS), soil temperature, and soil C content. The N_2O response to increasing N rate varies by region, indicating the importance of region-specific approaches for quantifying N_2O emissions and mitigation potential (Decock, 2014). These conclusions have led to the observations that there are differences among cropping systems, especially those which include cover crop as an effective strategy to reduce N_2O emissions because of the change in the soil water content and change in the soil NO_3^--N levels (Bayer et al., 2015; Chirinda et al., 2010; Li et al., 2015; Sauer et al., 2009; van der Weerden et al., 2000). Use of manure as a N source has produced mixed results on reduction of N_2O emissions, with some studies showing a positive effect on N_2O emission (e.g., Dittert et al., 2005; Hayakawa et al., 2009; Lampe et al., 2006; Plaza-Bonilla et al., 2014; Singurindy et al., 2009; Snyder et al., 2009; VanderZaag et al., 2011), while others showed a negative effect (Sauer et al., 2009). The negative effects of manure may be attributed to the tillage operations before and following manure applications, which increases the N_2O emissions (Singurindy et al., 2009).

Tillage systems interact with the N dynamics in soil, and the reduction in tillage can lead to a reduction in N_2O emissions as observed in the meta-analysis by Van Kessel et al. (2013), which found reduced tillage to be an advantage. A potential management strategy for the reduction of N_2O emissions is the alteration of the N form through the use of stabilizers or slow-release coatings; however, these studies have produced mixed results, with the more positive results in regions where soil water is managed (Halvorson et al., 2010a, 2010b; Halvorson and Del Grosso, 2012; Halvorson et al., 2014; Scheer et al., 2014; Snyder et al., 2009; Wang and Dalal, 2015; Zhang et al., 2015b) and less in the rain-fed areas where soil water content varies (Asgedom et al., 2013; Li et al., 2012; Maharjan et al., 2014; Parkin and Hatfield, 2014; Sistani et al., 2011; Venterea et al., 2005, 2011). The varying responses among these studies can be attributed to the length of time the observations were made, because the effect of the inhibitors or coatings are very detectable immediately after application but when considered for a whole year the effects are less obvious. Bhatia et al. (2010) compared N_2O emissions under conventional and no-till wheat with two nitrification inhibitors, S-benzylisothiouronium butanoate (SBTbutanoate) and S-benzylisothiouronium furoate (SBT-furoate). In this study, cumulative emission of N_2O-N was higher under no-tillage compared with conventional tillage by 12.2% for

urea fertilization and higher by 4.1 to 4.8% for the inhibitor treatments in the no-till compared with conventional tillage. However, the use of the nitrification inhibitors reduced the N_2O loss by nearly 50% of the applied N. Decock (2014) stated that the urease inhibitor N-(n-butyl) thiophosphoric triamide (NBPT) in combination with the nitrification inhibitor dicyandiamide (DCD) was the only management strategy that consistently reduced N_2O emissions, but the number of observations underlying this effect was relatively low.

Achieving a reduction in N_2O emissions from agricultural systems will require an integrated approach to develop regional strategies with the greatest chance of success (Snyder et al., 2009; VanderZaag et al., 2011; Nayak et al., 2015; Zhao et al., 2015). Zhao et al. (2015) proposed that the most effective method to decrease N_2O emissions would be to reduce the N surplus in the soil by increasing N use by the crop and optimizing N management. There is no consistent effect of any of the practices as summarized in Table 1 along with the factors which potentially interact to affect the overall response. The only consistent effect from any of the practices is N rate, and it is difficult to place an exact magnitude on the effect. The caveats surrounding each practice prevent a simple approach from being defined as the most effective solution to achieve reductions in N_2O emissions. An alternative will be to compare cropping system yield-scaled N_2O emissions which often show the combined set of practices implemented to

Table 1. Effect of different practices on reducing N_2O emissions from agricultural systems.

Practice	Effect	Interacting factors
Nitrogen rate	+	tillage, N inhibitors, crop rotation, soil C content, timing of application
N source	Neutral	
Placement	+	form, Rate, N timing, meteorological conditions, soil water content, soil C content
	−	form, rate, timing, soil water content at surface
Timing	+	form, rate, placement, meteorological conditions, soil water content, soil C content
Nitrification inhibitors	+	rate, N timing, Irrigation, meteorological conditions, soil C content
	−	N Timing, meteorological conditions
Tillage	+	N timing, Soil aggregation, Soil gas exchange, soil C content
	−	Soil water, N timing
Crop rotation	+	tillage, Precipitation pattern relative to N application, N inhibitors,
	Neutral	tillage, C content in soil
Soil water content	−	soil aggregation (gas exchange), Precipitation timing

achieve a greater production, with no change in N_2O emissions leading to the conclusion that the practice was not effective because the N_2O flux was not affected. There is a need for more data collection focused on side-by-side comparisons of management systems throughout the entire year to capture the dynamics of emissions during all of the cultural operations instead of just a single component. Improvements in biogeochemical simulation models which can be employed to help evaluate alternative strategies will provide insights into the interactions among the physical, biological, and chemical components of agricultural systems. We can implement practices which reduce N_2O emissions from agricultural systems, we just need to realize that there are several different pathways to achieve this result.

References

Aguilera, E., L. Lassaletta, A. Sanz-Cobena, J. Garnier, and A. Vallejo. 2013. The potential of organic fertilizers and water management to reduce N_2O emissions in Mediterranean climate cropping systems. A review. Agric. Ecosyst. Environ. 164:32–52. doi:10.1016/j.agee.2012.09.006

AAPFCO. 2013. Official publication number 65. Association of American Plant Food Control Officials. http://www.aapfco.org/publications.html (accessed 2 Dec. 2013).

Asgedom, H., M. Tenuta, D.N. Flaten, X. Gao, and E. Kebreab. 2013. Nitrous oxide emissions from a clay soil receiving granular urea formulations and dairy manure. Agron. J. 106:732–744. doi:10.2134/agronj2013.0096

Aulakh, M.S., D.A. Rennie, and E.A. Paul. 1984. Gaseous nitrogen losses from soils under zero-till as compared with conventional-till management-systems. J. Environ. Qual. 13:130–136. doi:10.2134/jeq1984.00472425001300010024x

Ball, B.C., B.S. Griffiths, C.F.E. Topp, R. Wheatley, R.L. Walker, R.M. Rees, C.A. Watson, H. Gordon, P.D. Hallett, B.M. McKenzie, and I.M. Nevison. 2014. Seasonal nitrous oxide emissions from field soils under reduced tillage, compost application or organic farming. Agric. Ecosyst. Environ. 189:171–180. doi:10.1016/j.agee.2014.03.038

Ball, B.C., A. Scott, and J.P. Parker. 1999. Field N_2O, CO_2 and CH_4 fluxes in relation to tillage, compaction and soil quality in Scotland. Soil Tillage Res. 53:29–39. doi:10.1016/S0167-1987(99)00074-4

Barton, L., R. Kiese, D. Gatter, K. Butterbach-Bahl, R. Buck, C. Hinz, and D.V. Murphy. 2008. Nitrous oxide emissions from a cropped soil in a semi-arid climate. Glob. Change Biol. 14:177–192.

Barton, L., D.V. Murphy, R. Kiese, and K. Butterbach-Bahl. 2010. Soil nitrous oxide and methane fluxes are low from a bioenergy crop (canola) grown in a semi-arid climate. Glob. Change Biol. Bioenerg. 2:1–15. doi:10.1111/j.1757-1707.2010.01034.x

Bateman, E.J., and E.M. Baggs. 2005. Contributions of nitrification and denitrification to N_2O emissions from soils at different water-filled pore space. Biol. Fertil. Soils 41:379–388. doi:10.1007/s00374-005-0858-3

Bayer, C., J. Gomes, J.A. Zanatta, F.C.B. Vieira, M. de Cássia Piccolo, J. Dieckow, and J. Six. 2015. Soil nitrous oxide emissions as affected by long-term tillage, cropping systems and nitrogen fertilization in Southern Brazil. Soil Tillage Res. 146:213–222. doi:10.1016/j.still.2014.10.011

Bhatia, A., S. Sasmal, N. Jain, H. Pathak, R. Kumar, and A. Singh. 2010. Mitigating nitrous oxide emission from soil under conventional and no-tillage in wheat using nitrification inhibitors. Agric. Ecosyst. Environ. 136:247–253. doi:10.1016/j.agee.2010.01.004

Bouwman, A.F., L.J.M. Boumans, and N.H. Batjes. 2002a. Emissions of N_2O and NO from fertilized fields: Summary of available measurement data. Glob. Biogeochem. Cycles 16:1058. doi:10:1029/2001GB001811

Bouwman, A.F., L.J.M. Boumans, and N.H. Batjes. 2002b. Modeling global annual N_2O and NO emissions from fertilized fields. Glob. Biogeochem. Cycles 16:1080. doi:10:1029/2001GB001812

Bremer, D.J. 2006. Nitrous oxide fluxes in turfgrass: Effects of nitrogen fertilization rates and types. J. Environ. Qual. 35:1678–1685. doi:10.2134/jeq2005.0387

Cai, Y., W. Ding, and J. Luo. 2012. Spatial variation of nitrous oxide emission between interrow soil and interrow plus row soil in a long-term maize cultivated sandy loam soil. Geoderma 181–182:2–10. doi:10.1016/j.geoderma.2012.03.005

Carter, J.P., Y.H. Hsiao, S. Spiro, and D.J. Richardson. 1995. Soil and sediment bacteria capable of aerobic nitrate respiration. Appl. Environ. Microbiol. 61:2852–2858.

Chirinda, N., M.S. Carter, K.R. Albert, P. Ambus, J.E. Olesen, J.R. Porter, and S.O. Petersen. 2010. Emissions of nitrous oxide from arable organic and conventional cropping systems on two soil types. Agric. Ecosyst. Environ. 136:199–208. doi:10.1016/j.agee.2009.11.012

Choudhary, M.A., A. Akramkhanov, and S. Saggar. 2001. Nitrous oxide emissions in soils cropped with maize under long-term tillage and under permanent pasture in New Zealand. Soil Tillage Res. 62:61–71. doi:10.1016/S0167-1987(01)00208-2

Choudhary, M.A., A. Akramkhanov, and S. Saggar. 2002. Nitrous oxide emissions from a New Zealand cropped soil: Tillage effects, spatial and seasonal variability. Agric. Ecosyst. Environ. 93:33–43. doi:10.1016/S0167-8809(02)00005-1

Cui, F., G. Yan, Z. Zhou, X. Zheng, and J. Deng. 2012. Annual emissions of nitrous oxide and nitric oxide from a wheat-maize cropping system on a silt loam calcareous soil in the North China Plain. Soil Biol. Biochem. 48:10–19. doi:10.1016/j.soilbio.2012.01.007

Decock, C. 2014. Mitigating nitrous oxide emissions from corn cropping systems in the Midwestern

U.S.: Potential and data gaps. Environ. Sci. Technol. 48:4247–4256. doi:10.1021/es4055324

De Antoni Migliorati, M., M. Bell, P.R. Grace, C. Scheer, D.W. Rowlings, and S. Liu. 2015. Legume pastures can reduce N₂O emissions intensity in subtropical cereal cropping systems. Agric. Ecosyst. Environ. 204:27–39. doi:10.1016/j.agee.2015.02.007

De Antoni Migliorati, M., C. Scheer, P.R. Grace, D.W. Rowlings, M. Bell, and J. McGree. 2014. Influence of different nitrogen rates and DMPP nitrification inhibitor on annual N₂O emissions from a subtropical wheat–maize cropping system. Agric. Ecosyst. Environ. 186:33–43.

De Klein, C., R.S.A. Novoa, S. Ogle, K.A. Smith, P. Rochette, and T.C. Wirth. 2006. 2006 IPPC Guidelines for national greenhouse gas inventories. Volume 4: Agriculture, forestry and other land use. Chapter 11. IGES, Japan. 54 pp.

Dittert, K., C. Lampe, R. Gasche, K. Butterbach-Bahl, M. Wachendorf, H. Papen, B. Sattelmacher, and F. Taube. 2005. Short-term effects of single or combined application of mineral N fertilizer and cattle slurry on the fluxes of radiatively active trace gases from grassland soil. Soil Biol. Biochem. 37:1665–1674. doi:10.1016/j.soilbio.2005.01.029

Dobbie, K.E., I.P. McTaggart, and K.A. Smith. 1999. Nitrous oxide emissions from intensive agricultural systems: Variations between crops and seasons, key driving variables, and mean emission factors. J. Geophys. Res. 104:26891–26899. doi:10.1029/1999JD900378

Dobbie, K.E., and K.A. Smith. 2003. Nitrous oxide emission factors for agricultural soils in Great Britain: The impact of soil water-filled pore space and other controlling variables. Glob. Change Biol. 9:204–218. doi:10.1046/j.1365-2486.2003.00563.x

Gogoi, B., and K.K. Baruah. 2012. Nitrous oxide emissions from fields with different wheat and rice varieties. Pedosphere 22:112–121. doi:10.1016/S1002-0160(11)60197-5

Granli, T. and O.C. Bøckman. 1994. Nitrous oxide from agriculture. Norw. J. Agric. Sci. Supplement No. 12, 1–128.

Grant, R.F., and E. Pattey. 2003. Modelling variability in N₂O emissions from fertilized agricultural fields. Soil Biol. Biochem. 35:225–243. doi:10.1016/S0038-0717(02)00256-0

Halvorson, A.D., and M.E. Bartolo. 2014. Nitrogen source and rate effects on irrigated corn yields and nitrogen-use efficiency. Agron. J. 106:681–693. doi:10.2134/agronj2013.0001

Halvorson, A.D., and S.J. Del Grosso. 2012. Nitrogen source and placement effects on soil nitrous oxide emissions from no-till corn. J. Environ. Qual. 41:1349–1360. doi:10.2134/jeq2012.0129

Halvorson, A.D., S.J. Del Grosso, and F. Alluvione. 2010a. Nitrogen source effects on nitrous oxide emissions from irrigated no-till corn. J. Environ. Qual. 39:1554–1562. doi:10.2134/jeq2010.0041

Halvorson, A.D., S.J. Del Grosso, and F. Alluvione. 2010b. Tillage and inorganic nitrogen source effects on nitrous oxide emissions from irrigated cropping systems. Soil Sci. Soc. Am. J. 74:436–445. doi:10.2136/sssaj2009.0072

Halvorson, A.D., C.S. Snyder, A.D. Blaylock, and S.J. Del Grosso. 2014. Enhanced-efficiency nitrogen fertilizers: Potential role in nitrous oxide emission mitigation. Agron. J. 106:715–722. doi:10.2134/agronj2013.0081

Harris, R.H., S.J. Officer, P.A. Hill, R.D. Armstrong, K.M. Fogarty, R.P. Zollinger, A.J. Phelan, and D.L. Partington. 2013. Can nitrogen fertiliser and nitrification inhibitor management influence N₂O losses from high rainfall cropping systems in South Eastern Australia? Nutr. Cycling Agroecosyst. 95:269–285. doi:10.1007/s10705-013-9562-0

Hayakawa, A., H. Akiyama, S. Sudo, and K. Yagi. 2009. N₂O and NO emissions from an Andisol field as influenced by pelleted poultry manure. Soil Biol. Biochem. 41:521–529. doi:10.1016/j.soilbio.2008.12.011

Horváth, L., B. Grosz, A. Machon, Z. Tuba, Z. Nagy, S.Z. Czóbel, J. Balogh, E. Péli, S.Z. Fóti, T. Weidinger, K. Pintér, and E. Führer. 2010. Estimation of nitrous oxide emission from Hungarian semi-arid sandy and loess grasslands: Effect of soil parameters, grazing, irrigation and use of fertilizer. Agric. Ecosyst. Environ. 139:255–263. doi:10.1016/j.agee.2010.08.011

Howard, D.M., and P.J.A. Howard. 1993. Relationships between CO₂ evolution, moisture content and temperature for a range of soil types. Soil Biol. Biochem. 25:1537–1546. doi:10.1016/0038-0717(93)90008-Y

Kim, D.-G., G. Hernandez-Ramirez, and D. Giltrap. 2013. Linear and nonlinear dependency of direct nitrous oxide emissions on fertilizer nitrogen input: A meta-analysis. Agric. Ecosyst. Environ. 168:53–65. doi:10.1016/j.agee.2012.02.021

Knowles, R. 1982. Denitrification. Microbiol. Rev. 46:43–70.

Kong, A.Y.Y., S.J. Fonte, C. van Kessel, and J. Six. 2009. Transitioning from standard to minimum tillage: Trade-offs between soil organic matter stabilization, nitrous oxide emissions, and N availability in irrigated cropping systems. Soil Tillage Res. 104:256–262. doi:10.1016/j.still.2009.03.004

Lampe, C., K. Dittert, B. Sattelmacher, M. Wachendorf, R. Loges, and F. Taube. 2006. Sources and rates of nitrous oxide emissions from grazed grassland after application of ¹⁵N-labelled mineral fertilizer and slurry. Soil Biol. Biochem. 38:2602–2613. doi:10.1016/j.soilbio.2006.03.016

Lesschen, J.P., G.L. Velthof, W. de Vries, and J. Kros. 2011. Differentiation of nitrous oxide emission factors for agricultural soils. Environ. Pollut. 159:3215–3222. doi:10.1016/j.envpol.2011.04.001

Li, C., X. Hao, R.E. Blackshaw, J.T. O'Donovan, K.N. Harker, and G.W. Clayton. 2012. Nitrous oxide emissions in response to ESN and urea, herbicide management and canola cultivar in a no-till cropping system. Soil Tillage Res. 118:97–106. doi:10.1016/j.still.2011.10.017

Li, X., S.O. Petersen, P. Sørensen, and J.E. Olesen. 2015. Effects of contrasting catch crops on nitrogen availability and nitrous oxide emissions in an organic cropping system. Agric. Ecosyst. Environ. 199:382–393. doi:10.1016/j.agee.2014.10.016

Liu, J., L. Zhu, S. Luo, L. Bu, X. Chen, S. Yue, and S. Li. 2014. Response of nitrous oxide emission to soil mulching and nitrogen fertilization in semiarid farmland. Agric. Ecosyst. Environ. 188:20–28. doi:10.1016/j.agee.2014.02.010

Lloyd, J., and A. Taylor. 1994. On the temperature dependence of soil respiration. Funct. Ecol. 8:315–323. doi:10.2307/2389824

Maharjan, B., and R.T. Venterea. 2013. Nitrite intensity explains N management effects on N_2O emissions in maize. Soil Biol. Biochem. 66:229–238. doi:10.1016/j.soilbio.2013.07.015

Maharjan, B., R.T. Venterea, and C. Rosen. 2014. Fertilizer and irrigation management effects on nitrous oxide emissions and nitrate leaching. Agron. J. 106:703–714. doi:10.2134/agronj2013.0179

Mahmood, T., R. Ali, K.A. Malik, and S.R.A. Shamsi. 1998. Nitrous oxide emissions from an irrigated sandy-clay loam cropped to maize and wheat. Biol. Fertil. Soils 27:189–196. doi:10.1007/s003740050419

Manalil, S., and K. Flower. 2014. Soil water conservation and nitrous oxide emissions from different crop sequences and fallow under Mediterranean conditions. Soil Tillage Res. 143:123–129. doi:10.1016/j.still.2014.06.006

Martins, M.R., C.P. Jantalia, J.C. Polidoro, J.N. Batista, B.J.R. Alves, R.M. Boddey, and S. Urquiaga. 2015. Nitrous oxide and ammonia emissions from N fertilization of maize crop under no-till in a Cerrado soil. Soil Tillage Res. 151:75–81. doi:10.1016/j.still.2015.03.004

Metivier, K.A., E. Pattey, and R.F. Grant. 2009. Using the *ecosys* mathematical model to simulate temporal variability of nitrous oxide emissions from a fertilized agricultural soil. Soil Biol. Biochem. 41:2370–2386. doi:10.1016/j.soilbio.2009.03.007

Mosier, A., C. Kroeze, C. Nevison, O. Oenema, S. Seitzinger, and O. van Cleemput. 1998. Closing the global N_2O budget: Nitrous oxide emissions through the agricultural nitrogen cycle. Nutr. Cycl. Agroecosyst. 52:225–248. doi:10.1023/A:1009740530221

Mu, Z., A. Huang, S.D. Kimura, T. Jin, S. Wei, and R. Hatano. 2009. Linking N_2O emission to soil mineral N as estimated by CO_2 emission and soil C/N ratio. Soil Biol. Biochem. 41:2593–2597. doi:10.1016/j.soilbio.2009.09.013

Nayak, D., E. Saetnan, K. Cheng, W. Wang, F. Koslowski, Y. Cheng, W.Y. Zhu, J. Wang, J. Liu, D. Moran, X. Yan, L. Cardenas, J. Newbold, G. Pan, Y. Lu, and P. Smith. 2015. Management opportunities to mitigate greenhouse gas emissions from Chinese agriculture. Agric. Ecosyst. Environ. 209:108–124. doi:10.1016/j.agee.2015.04.035

Parkin, T.B., and J.L. Hatfield. 2014. Enhanced efficiency fertilizers: Effect on nitrous oxide emissions in Iowa. Agron. J. 106:694–702. doi:10.2134/agronj2013.0219

Patureau, D., E. Zumstein, J.P. Delgenes, and R. Moletta. 2000. Aerobic denitrifiers isolated from diverse natural and managed ecosystems. Microb. Ecol. 39:145–152. doi:10.1007/s002480000009

Pelster, D.E., F. Larouche, P. Rochette, M.H. Chantigny, S. Allaire, and D.A. Angers. 2011. Nitrogen fertilization but not soil tillage affects nitrous oxide emissions from a clay loam soil under a maize–soybean rotation. Soil Tillage Res. 115-116:16–26. doi:10.1016/j.still.2011.06.001

Petersen, S.O., P. Schjønning, I.K. Thomsen, and B.T. Christensen. 2008. Nitrous oxide evolution from structurally intact soil as influenced by tillage and soil water content. Soil Biol. Biochem. 40:967–977. doi:10.1016/j.soilbio.2007.11.017

Pfab, H., I. Palmer, F. Buegger, S. Fiedler, T. Muller, and R. Ruser. 2012. Influence of a nitrification inhibitor and of placed N-fertilization on N_2O fluxes from a vegetable cropped loamy soil. Agric. Ecosyst. Environ. 150:91–101. doi:10.1016/j.agee.2012.01.001

Plaza-Bonilla, D., J. Álvaro-Fuentes, J. Luis Arrúe, and C. Cantero-Martínez. 2014. Tillage and nitrogen fertilization effects on nitrous oxide yield-scaled emissions in a rainfed Mediterranean area. Agric. Ecosyst. Environ. 189:43–52. doi:10.1016/j.agee.2014.03.023

Sánchez-Martín, L., A. Vallejo, J. Dick, and U.M. Skiba. 2008a. The influence of soluble carbon and fertilizer nitrogen on nitric oxide and nitrous oxide emissions from two contrasting agricultural soils. Soil Biol. Biochem. 40:142–151. doi:10.1016/j.soilbio.2007.07.016

Sánchez-Martín, L., A. Arce, A. Benito, L. Garcia-Torres, and A. Vallejo. 2008b. Influence of drip and furrow irrigation systems on nitrogen oxide emissions from a horticultural crop. Soil Biol. Biochem. 40:1698–1706. doi:10.1016/j.soilbio.2008.02.005

Sauer, T.J., S.R. Compston, C.P. West, G. Hernandez-Ramirez, E.E. Gbur, and T.B. Parkin. 2009. Nitrous oxide emissions from a bermudagrass pasture: Interseeded winter rye and poultry litter. Soil Biol. Biochem. 41:1417–1424. doi:10.1016/j.soilbio.2009.03.019

Sander, B.O., M. Samson, and R.J. Buresh. 2014. Methane and nitrous oxide emissions from flooded rice fields as affected by water and straw management between rice crops. Geoderma 235–236:355–362. doi:10.1016/j.geoderma.2014.07.020

Scheer, C., D.W. Rowlings, M. Firrel, P. Deuter, S. Morris, and P.R. Grace. 2014. Impact of nitrification inhibitor (DMPP) on soil nitrous oxide emissions from an intensive broccoli production system in sub-tropical Australia. Soil Biol. Biochem. 77:243–251. doi:10.1016/j.soilbio.2014.07.006

Scheer, C., R. Wassmann, K. Kienzler, N. Ibragimov, and R. Eschanov. 2008. Nitrous oxide emissions from fertilized, irrigated cotton (*Gossypium hirsutum* L.) in the Aral Sea Basin, Uzbekistan: Influence of nitrogen applications and irrigation practices. Soil Biol. Biochem. 40:290–301. doi:10.1016/j.soilbio.2007.08.007

Sheehy, J., J. Six, L. Alakukku, and K. Regina. 2013. Fluxes of nitrous oxide in tilled and no-tilled boreal arable soils. Agric. Ecosyst. Environ. 164:190–199. doi:10.1016/j.agee.2012.10.007

Shoun, H., D.-H. Kim, H. Uchiyama, and J. Sugiyama. 1992. Denitrification by fungi. FEMS Microbiol. Lett. 94:277–281. doi:10.1111/j.1574-6968.1992.tb05331.x

Singurindy, O., M. Molodovskaya, B.K. Richards, and T.S. Steenhuis. 2009. Nitrous oxide emission at low temperatures from manure-amended soils under corn (*Zea mays* L.). Agric. Ecosyst. Environ. 132:74–81. doi:10.1016/j.agee.2009.03.001

Sistani, K.R., M. Jn-Baptiste, N. Lovanh, and K.L. Cook. 2011. Atmospheric emissions of nitrous oxide, methane, and carbon dioxide from different nitrogen fertilizers. J. Environ. Qual. 40:1797–1805. doi:10.2134/jeq2011.0197

Smith, K., D. Watts, T. Way, H. Torbert, and S. Prior. 2012. Impact of tillage and fertilizer application method on gas emissions in a corn cropping system. Pedosphere 22:604–615. doi:10.1016/S1002-0160(12)60045-9

Smith, P., K.W. Goulding, K.A. Smith, D.S. Powlson, J.U. Smith, P. Falloon, and K. Coleman. 2001. Enhancing the carbon sink in European agricultural soils: Including trace gas fluxes in estimates of carbon mitigation potential. Nutr. Cycling Agroecosyst. 60:237–252. doi:10.1023/A:1012617517839

Snyder, C.S., T.W. Bruulsema, T.L. Jensen, and P.E. Fixen. 2009. Review of greenhouse gas emissions from crop production systems and fertilizer management effects. Agric. Ecosyst. Environ. 133:247–266. doi:10.1016/j.agee.2009.04.021

Tan, I.Y.S., H.M. van Es, J.M. Duxbury, J.J. Melkonian, R.R. Schindelbeck, L.D. Geohring, W.D. Hively, and B.N. Moebius. 2009. Single-event nitrous oxide losses under maize production as affected by soil type, tillage, rotation, and fertilization. Soil Tillage Res. 102:19–26. doi:10.1016/j.still.2008.06.005

USEPA. 2014. Inventory of US Greenhouse Gases and Sinks: 1990–2012. EPA 430-R-14-003. Washington, DC. 529 pp. www.epa.gov/climatechange/ghgemissions/usinventoryreport (accessed 27 Jan. 2015).

Ussiri, D.A.N., R. Lal, and M.K. Jarecki. 2009. Nitrous oxide and methane emissions from long-term tillage under a continuous corn cropping system in Ohio. Soil Tillage Res. 104:247–255. doi:10.1016/j.still.2009.03.001

Uzoma, K.C., W. Smith, B. Grant, R.L. Desjardins, X. Gao, K. Hanis, M. Tenuta, P. Goglio, and C. Li. 2015. Assessing the effects of agricultural management on nitrous oxide emissions using flux measurements and the DNDC model. Agric. Ecosyst. Environ. 206:71–83. doi:10.1016/j.agee.2015.03.014

Van Kessel, C., R. Venterea, J. Six, M.A. Adviento-Borbe, B. Linquist, and K.J. VanGroenigen. 2013. Climate, duration, and N placement determine N_2O emissions in reduced tillage systems: A meta-analysis. Glob. Change Biol. 19:33–44. doi:10.1111/j.1365-2486.2012.02779.x

van der Weerden, T.J., R.R. Sherlock, P.H. Williams, and K.C. Cameron. 2000. Effect of three contrasting onion (*Allium cepa* L.) production systems on nitrous oxide emissions from soil. Biol. Fertil. Soils 31:334–342. doi:10.1007/s003740050665

VanderZaag, A.C., S. Jayasundara, and C. Wagner-Riddle. 2011. Strategies to mitigate nitrous oxide emissions from land applied manure. Anim. Feed Sci. Technol. 166-167:464–479. doi:10.1016/j.anifeedsci.2011.04.034

Venterea, R.T., M. Burger, and K.A. Spokas. 2005. Nitrogen oxide and methane emissions under varying tillage and fertilizer management. J. Environ. Qual. 34:1467–1477. doi:10.2134/jeq2005.0018

Venterea, R.T., B. Maharjan, and M.S. Dolan. 2011. Fertilizer source and tillage effects on yield-scaled nitrous oxide emissions in a corn cropping system. J. Environ. Qual. 40:1521–1531. doi:10.2134/jeq2011.0039

Wang, W., and R.C. Dalal. 2015. Nitrogen management is the key for low-emission wheat production in Australia: A life cycle perspective. Eur. J. Agron. 66:74–82. doi:10.1016/j.eja.2015.02.007

Welzmiller, J.T., A.D. Matthias, S. White, and T.L. Thompson. 2008. Elevated carbon dioxide and irrigation effects on soil nitrogen gas exchange in irrigated sorghum. Soil Sci. Soc. Am. J. 72:393–401. doi:10.2136/sssaj2007.0033

Zhang, M., C.H. Fan, Q.L. Li, B. Li, Y.Y. Zhu, and Z.Q. Xiong. 2015a. A 2-yr field assessment of the effects of chemical and biological nitrification inhibitors on nitrous oxide emissions and nitrogen use efficiency in an intensively managed vegetable cropping system. Agric. Ecosyst. Environ. 201:43–50. doi:10.1016/j.agee.2014.12.003

Zhang, Y., J. Sheng, Z. Wang, L. Chen, and J. Zheng. 2015b. Nitrous oxide and methane emissions from a Chinese wheat–rice cropping system under different tillage practices during the wheat-growing season. Soil Tillage Res. 146:261–269. doi:10.1016/j.still.2014.09.019

Zhao, M., Y. Tian, Y. Mac, M. Zhang, Y. Yao, Z. Xiong, B. Yin, and Z. Zhu. 2015. Mitigating gaseous nitrogen emissions intensity from a Chinese rice cropping system through an improved management practice aimed to close the yield gap. Agric. Ecosyst. Environ. 203:36–45. doi:10.1016/j.agee.2015.01.014

Zhu-Barker, Z., W.R. Horwath, and M. Burger. 2015. Knife-injected anhydrous ammonia increases yield-scaled N2O emissions compared to broadcast or band-applied ammonium sulfate in wheat. Agric. Ecosyst. Environ. 212:148–157. doi:10.1016/j.agee.2015.06.025

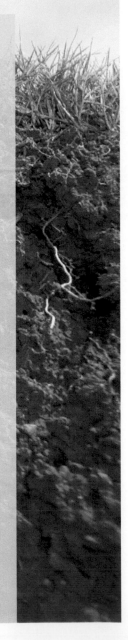

The Use of Enriched and Natural Abundance Nitrogen and Carbon Isotopes in Soil Fertility Research

David Clay,* Cheryl Reese, Stephanie A.H. Bruggeman, and Janet Moriles-Miller

Providing food for a rapidly growing world population relies on (i) increasing amount of food produced on cultivated land, (ii) increasing the genetic yield potentials, (iii) improving food transportation, and (iv) reducing food losses. This chapter addresses techniques that can be used to increase productivity on currently cultivated land and it provides a review of stable isotope applications in ecological research, examples on how to convert isotopic values from one unit to another, demonstrates how to use isotopic analysis to solve complex problems, and provides examples on how to determine a N fertilizer efficiency by using ^{15}N-enriched and depleted fertilizers.

Stable Isotopes Use in Ecological Systems

Natural abundance and enriched ^{13}C and ^{15}N isotopes can be used for many different purposes that include: (i) determining soil organic carbon (SOC) turnover; (ii) assessing N fertilizer dynamics and soil health (Clapp et al., 2000; Clay et al., 2005; Mamani-Pati et al., 2010); (iii) determining animal migration patterns and trophic level (Tiunov, 2007; Harding and Hart, 2011); (iv) assessing nitrogen, carbon, and water transport in watersheds (Kendal, 1989; Hay, 2004; Clay et al., 2007); (v) assessing the value of cover crops (Smeltekop et al., 2002; Reese et al., 2014); (vi) determining water stress in sap flow (Pate and Arthur, 1998); and (vii) assessing factors limiting crop growth and quality (Jefferies and Mackerron, 1997; Kharel et al., 2011; Park et al., 2014). Even though slightly different methods are required for each problem, the basic concepts across problems are identical. The methods are based on two concepts. First, the isotopic approach provides information needed to mathematically solve problems that previously were unsolvable. For example,

---

Abbreviations: At%, atom percent; DHO, deuterium-labeled water; NHC, nonharvested carbon; PCR, plant carbon retained; PDB, Pee Dee Belemnite; SOC, soil organic carbon.

David Clay, Cheryl Reese (Cheryl.Reese@sdstate.edu), Stephanie A.H. Bruggeman (Stephanie.Bruggeman@sdstate.edu), and Janet Moriles-Miller (janet.miller@sdstate.edu), Plant Science Dep., South Dakota State Univ., Berg Agricultural Hall, P.O. Box 2207A, Brookings, SD 57007. *Corresponding author (david.clay@sdstate.edu).

doi:10.2134/soilfertility.2014.0013

Soil Fertility Management in Agroecosystems
Amitava Chatterjee and David Clay, Editors

solving equations with two or three unknowns requires an equation for each unknown. Isotopic analysis provides those equations (Kim et al., 2008; Clay et al., 2009). Second, the values by themselves have value. For example, soil with a soil organic C δ^{13}C value of –22‰ suggest that the primary source of plant material to the SOC was from a C_3 plant. In a second example, Jahren and Kraft (2008) investigated the isotopic signatures of fast foods. They reported that McDonalds beef had a δ^{13}C value of –17.7‰, whereas Burger King beef had a δ^{13}C value of –20.8‰. Differences in the values suggest that the diets of beef used to produce McDonalds and Burger King's hamburgers were different.

Stable Isotopes and Nitrogen Fertilizer Efficiency

The traditional approach to assess nutrient stress was to determine the nutrient uptake efficiency. In this approach, the N use efficiency or the yield per unit fertilizer can be calculated by the equations

$$\text{N use efficiency} = \left(\frac{\text{N uptake}_{\text{fertilized}} - \text{N uptake}_{0\ \text{N control}}}{\text{N applied}} \right) \quad [1]$$

$$\text{Yield per unit fertilizer} = \left(\frac{\text{N yield}_{\text{fertilized}} - \text{N yield}_{0\ \text{N control}}}{\text{N applied}} \right) \quad [2]$$

In Eq. [1 and 2], the entire yield response is attributed to the added fertilizer. This function defines the N response for each treatment and does not consider synergistic relationships between N and water (Kim et al., 2008; Clay et al., 2009).

Even though numerous studies have shown that water has synergistic and antagonistic impacts on N availability, most N recommendation models do not directly consider water availability in the calculation. For example, one of the most commonly used models to calculate fertilizer recommendations is NR = k (yield goal) – N credits, where, NR is the recommended N rate, k ranges 21.4 to 26.8 kg N Mg^{-1} corn grain, N credits include soil NO_3–N provided by legumes, N in irrigation water, and NR is kg N ha^{-1} (Kim et al., 2013). Modifications of this basic model were used to design models for other crops. Basic limitations of this model were that (i) it does not consider water

as a nutrient, (ii) it does not consider N mineralization, (iii) it does not consider the impact of water on nutrient flux, and (iv) synergistic relationships between water and the nutrient are not considered (Kim et al., 2008). This model has had mixed successes in site-specific applications (Fox and Piskielek, 1995; Lory and Scharf, 2003).

Based on ^{15}N and ^{13}C stabile isotope analysis, Kim et al. (2008) developed a conceptual model that integrated water into the decision process. The basis of this model was that fertilizer efficiency increases with decreasing water stress. In the NR model above, this could be accomplished by replacing k with k_f where f is a function of water and fertilizer application approach. As shown by Kim et al. (2008) k_r would increase with water. The mathematics behind this solution was built on solving the equation, yield = f(water, nitrogen). The power of the isotopic approach was that it provides independent equations needed to solve complex relationships.

Stable Isotopic Review

Multiple stable isotopes are the result of atoms containing identical numbers of protons and electrons but different numbers of neutrons (Table 1). For example, ^{13}C has seven neutrons and six protons, whereas ^{12}C has six protons and six neutrons. Stable isotopes have the added benefits in that they are not radioactive and do not require special use permits. Some elements, such as ^{32}P, contain only one stable isotope, whereas other elements contain multiple isotopes (Table 1).

Stable isotopes are used in ecological source-tracking experiments because elements with different mass weights react at different rates. For example, deuterium-labeled water (DHO) evaporates slower than H_2O, and therefore its concentration increases during evaporation. The amount of fractionation depends location. In areas with high evaporation, the DHO/H_2O ratio will be higher than in areas with low evaporation. Isotopic fractionation produces an isotopic signal that can be tracked, and the strength of the signal is dependent on the isotopic fractionation and the length of time that an animal ate or drank the labeled material. On the basis of isotopic fractionation, duck feathers collected in Brookings, SD, can be used to help identify the bird's migration pattern (Henaux et al., 2012).

Table 1. Stable isotopes commonly used in soil fertility research.

Isotope	%	Isotope	%	Isotope	%
H	99.9844	^{16}O	99.763	^{39}K	93.08
D	0.0156	^{17}O	0.0375	^{40}K	0.0119
^{12}C	98.89	^{18}O	0.1995	^{41}K	6.91
^{13}C	1.11	^{32}S	95.02		
^{14}N	99.64	^{33}S	0.75		
^{15}N	0.36	^{34}S	4.21		

Stable Isotope Definitions

To maximize confusion, different disciplines use different approaches to describe isotopic compositions. The units for isotopic enriched tracer experiments are reported in atom percent (At%). The atom percent is the absolute number of atoms of a given isotope in 100 atoms of the total element. The definition for At% and At% excess are shown in Eq. [3].

$$At\%^{15}N = \left(\frac{^{15}N}{^{14}N + ^{15}N}\right)(100At\%)$$

$$At\%^{14}N = \left(\frac{^{14}N}{^{14}N + ^{15}N}\right)(100At\%) \quad [3]$$

$$100\% = At\%^{15}N + At\%^{14}N$$

$$At\%^{15}N \text{ excess} =$$
$$At\%^{15}N - \text{background } ^{15}N \text{ value}$$

The At%^{15}N of air is approximately 0.366%, and the At%^{15}N excess of air is 0%.

Similar calculations are used to define ^{13}C enrichments:

$$At\%^{13}C = \left(\frac{^{13}C}{^{12}C + ^{13}C + ^{14}C}\right)(100At\%)$$

$$At\%^{12}C = \left(\frac{^{12}C}{^{12}C + ^{13}C + ^{14}C}\right)(100At\%) \quad [4]$$

$$100 = At\%^{12}C + At\%^{13}C + At\%^{14}C$$

In Eq. [4], ^{14}C occurs in trace amount and represents approximately 1 part per trillion of $^{13}C + ^{12}C$. Because of low concentration, At%^{14}C is assumed to be approximately zero. If At%^{14}C equals 0, then $100 = At\%^{13}C + At\%^{12}C$. The primary problem with the At% value is that it is difficult to show small isotopic differences. For example, are two samples that have At%^{13}C values of 1.11 and 1.10% different? Based on initial

inspection there is a 0.01% difference between the two samples, and therefore the two values are similar. However, most ratio mass spectrometers have the capacity to detect these differences (Clay et al., 2007). For example, Clay reported that the standard deviation for the $\delta^{13}C$ value of wheat grain ($\delta^{13}C = -24.64‰$) much less than 1‰. The ^{13}C At% of $-25.5‰$ is 1.0832, whereas the ^{13}C At% of $-23.5‰$ is 1.08540. Differences between these values are in the fourth significant figure.

To reduce the number of reported significant figures, isotopic values can be reported as delta (δ) or ^{13}C isotopic discrimination values (Δ). These values are defined below, where R_N and R_C represent the atmospheric Ratios of N and C, respectively.

$$R_N = \left(\frac{At\%^{15}N}{At\%^{14}N}\right)$$

$$\delta^{15}N = \left(\frac{R_{N\text{ sample}} - R_{N\text{ std}}}{R_{N\text{ std}}}\right)(1000‰) \quad [5]$$

$$R_C = \left(\frac{At\%^{13}C}{At\%^{12}C}\right)$$

$$\delta^{13}C = \left(\frac{R_{C\text{ sample}} - R_{C\text{ std}}}{R_{C\text{ std}}}\right)(1000‰) \quad [6]$$

The δ values are reported in parts per thousand or per mil and represent the difference between the sample R and the standard R value. For N, the standard reference source is the N contained in air, whereas for C the standard reference materials is Pee Dee Belemnite (PDB). When reporting delta values, the source of the primary standard must be provided. Unfortunately, PDB reference material is not available and therefore secondary are used. Selected primary standards can be obtained from the

International Atomic Energy Agency (2013) and R values for the standards are provided below.

Deuterium–Hydrogen Standard Reference

Standard Antarctic precipitation (SLAP) $R_{D,SLAP}$ = 0.000089089

Vienna standard mean ocean water (VSMOW) $R_{D,VSMOW}$ = 0.00015575

^{15}N standard reference (air)

At%^{15}N = 0.3663033 and R_N = 0.0036436, whereas

^{13}C standard reference (PDB)

At%^{13}C = 1.1112328l and R_c = 0.0112372.

^{18}O standard primary references

SLAP $R_{O,SLAP}$ = 0.018939

VSMOW $R_{O,VSMOW}$ = 0.0020052

PDB $R_{O,PDB}$ = 0.0020672

In most situations, the $\delta^{13}C$ values are negative and often range from –8 to –25‰. The relative ^{13}C amount decreases with decreasing $\delta^{13}C$ value. To eliminate the negative values and make the values relative to the atmosphere, the term ^{13}C isotopic discrimination (Δ) was defined. This value is calculated with the equation,

$$\Delta = \left(\frac{\delta^{13}C_{PDB \ of \ CO2 \ in \ air} - \delta^{13}C_{PDBsample}}{1 + \delta^{13}C_{PDBsample}/1000} \right) \quad [7]$$

The new delta value (Δ) is relative to air, not limestone. A value of 0.0‰ indicates that air and the sample have identical relative amounts of ^{13}C. A value of 10‰ indicates that the sample is depleted in ^{13}C. A Δ value of 0 indicates that ^{13}C discrimination did not occur, whereas positive values indicate than ^{13}C discrimination occurred. In Eq. [7], $\delta^{13}_{CPDB \ of \ CO2 \ in \ air}$ has a value of –8‰ and a corresponding R_C value of 0.01115. As with ^{15}N, the δ and Δ values are reported in parts per thousand, or per mil (‰). Delta values are not absolute isotope abundances but differences between sample readings and a standard. The problems below provide examples on how to convert from one unit to another.

Problem 1. Determine the At%^{13}C if the $\delta^{13}C$ value is –19‰.

1. $\delta^{13}C = \left(\dfrac{R_{C \ sample} - R_{C \ std}}{R_{C \ std}} \right)(1000‰)$

$\delta^{13}C = \left(\dfrac{R_{C \ sample} - 0.0112372}{0.0112372} \right)(1000‰)$

$= -19‰$

$R_{C \ sample} = 0.01102369$

2. $100 = At\%^{12}C + At\%^{13}C + At\%^{14}C$

$At\%^{13}C = \left(\dfrac{^{13}C}{^{12}C + ^{13}C + ^{14}C} \right)(100 At\%)$

$At\%^{12}C = \left(\dfrac{^{12}C}{^{12}C + ^{13}C + ^{14}C} \right)(100 At\%)$

assume that $+ At\%^{14}C \approx 0$

so, $100 = At\%^{12}C + At\%^{13}C$, or

$100 - At\%^{13}C = At\%^{12}C$

3. $R_{C \ sample} = \left(\dfrac{At\%^{13}C}{At\%^{12}C} \right)$

$R_{C \ sample} = 0.01102369 = \dfrac{At\%^{13}C}{(100 - At\%^{13}C)}$

$At\%^{13}C = 1.090349$

Conducting this analysis is time consuming and difficult. Steps 1, 2, and 3 can be simplified by using the equation

$$At\%^{13}C = \frac{[100 R_{C \ std}] \left[\dfrac{\delta^{13}C}{1000} + 1 \right]}{1 + [R_{C \ std}] \left[\dfrac{\delta^{13}C}{1000} + 1 \right]} \quad [8]$$

To test this equation, the At% ^{13}C values from the two approaches are compared.

$$At\%^{13}C = \frac{[100 \times 0.0112372] \left[\dfrac{-19}{1000} + 1 \right]}{1 + [0.0112372] \left[\dfrac{-19}{1000} + 1 \right]}$$

$$At\%^{13}C = \frac{1.102369}{1.011024} = 1.09035 \quad [9]$$

Problem 2. A sample has a $\delta^{13}C$ value of $-16‰$; what is the Δ value (^{13}C isotopic discrimination)?

$$\Delta = \left(\frac{\delta^{13}C_{air} - \delta^{13}C_{sample}}{1 + \frac{\delta^{13}C_{sample}}{1000}} \right)$$

$$\Delta = \left(\frac{-8 - (-16)}{1 + \frac{(-16)}{1000}} \right) = 8.13$$

Based on this calculation the Δ value is 8.13‰.

Problem 3. A sample has an R_N value of 0.003964; what is its $\delta^{15}N$ value?

$$R_{N,standard} = 0.0036436$$

$$\delta^{15}N = \left(\frac{R_{N\ sample} - R_{N\ std}}{R_{N\ std}} \right)(1000‰)$$

$$\delta^{15}N = \left(\frac{0.003964 - 0.0036436}{0.0036436} \right)1000 = 87.94$$

Based on this calculation the $\delta^{15}N$ value is 87.94‰.

Calculating Fertilizer Responses

Based on the measured amounts of stable isotopes contained in a sample, the N fertilizer contribution to total N in the plant can be determined. For example, the contribution of N from different sources can be calculated with a mixing equation demonstrated in Problem 4. This approach is based on fertilizer and soil having different N signatures. For example, N derived from fertilizer typically or N fixation has a $\delta^{15}N$ value <0, while the soil-derived N from organic matter decomposition has $\delta^{15}N$ values ranging from 3 to 5‰. Soil-derived N has a higher $\delta^{15}N$ value than fertilizer because ^{15}N accumulates as soil organic matter as it decomposes. Based on these differences, the amount of fertilizer contained in the plant tissue can be calculated (Kim et al., 2008).

Problem 4. If the fertilizer has a $\delta^{15}N$ value of $-1.5‰$, harvested corn grain has a $\delta^{15}N$ value of 2.6‰, and corn from unfertilized areas has a $\delta^{15}N$ value of 6‰. What is the percentage of corn derived from fertilizer?

$$\% \text{ from soil} = \frac{(sample - unfert)}{(fertilizer - unfert)}100\%$$

$$= \frac{(2.6 - 6)}{(-1.5 - 6)}100\% = \frac{-3.5}{-7.5}100 = 46.7\%$$

$$\% \text{ from fertilizer} = 100 - 46.7\% = 53.3\%$$

Understanding Crop Limiting Factors

Yield can be defined by the equation, Yield = f (water, nutrients, light, diseases, insects, plus a multitude of other factors). Of course, this function is impossible to solve. Therefore, experiments attempt to control and/or eliminate factors seen as extraneous to the study. In the northern Great Plains, water and N are the most limiting factors, and therefore these nutrients are the focal points of many experiments. In these experiments, the generalized yield loss equation can be simplified into the form:

Measured yield = optimum yield
– yield loss due to N and water stress   [10]

In this equation, factors not specifically influenced by the imposed treatments and are maintained through management at a nonlimiting level. To develop a mechanistic understanding on how N and water interact to affect yield, the N and water factors need to be separated (Clay et al., 2001a, 2001b). These factors can be separated by using ^{13}C isotopic fractionation that occurs during photosynthesis. When water is not limiting, stomata are open and CO_2 diffuses freely in and out of the leaf. As water stress limits growth, plants reduce water loss by closing the stomates. This closure reduces CO_2 exchange between the leaf interior and the atmosphere, and in C_4 plants, results in more negative $\delta^{13}C$ values, and in C_3, less negative values. Nitrogen influences ^{13}C fractionation by impacting the amount of chlorophyll contained in the leaf.

Water stress changes plants' Δ value because carbon fixation is influenced by water availability. Atmospheric CO_2 contains 98.89% $^{12}CO_2$ and 1.11% $^{13}CO_2$. In C_3 plants, RuBisCo (ribulose bisphosphate carboxylase) discriminates against ^{13}C and water stress reduces this discrimination. When C_3 plants are not water stressed and gas exchange is taking place through the open stomates, the amount of ^{13}C fixed into sugar is very low. Under these conditions, RuBisCo preferentially uses $^{12}CO_2$ during the photosynthesis reaction. In response to water stress, plants close their stomata, and the relative amount of ^{13}C in the substomata air chamber increases, which in turn increases the fixation of $^{13}CO_2$. When the stomates are closed, δ^{13}C increases, (-26‰ → -22‰).

Research conducted by Carelli et al. (1999) showed that Δ is also influenced by N stress. Clay et al. (2001a) confirmed these results in a C_3 plant, and they showed that N and water stress had opposite effects on Δ, with N stress increasing carbon discrimination and water stress decreasing discrimination. The net result was that plants undergoing both N and water stress had identical Δ values as plants that were not undergoing N and water stress.

Separating Nitrogen and Water Stress

Experiments conducted in the Nebraska panhandle, north central Montana, and east central South Dakota (Clay et al., 2001a, 2003, 2005), showed that both N and water impacted yield and Δ. Based on this research, the following equation was developed:

$$\delta\Delta = \text{YLWS}(\delta\Delta_{\text{water stress}}) + \text{YLNS}(\delta\Delta_{\text{N stress}}) \quad [11]$$

where YLWS is yield loss due to water stress, YLNS is yield loss due to N stress, $\delta\Delta_{\text{water stress}}$ is change in Δ‰ as influenced by water stress, and $\delta\Delta_{\text{N stress}}$ is change in Δ‰ as influenced by N stress. Solving Eq. [10 and 11] requires the measurement of the $\delta\Delta\text{water}_{\text{stress}}$ and $\delta\Delta_{\text{N stress}}$ values. These values are determined in experiments that contain the appropriate control treatments. The $\delta\Delta_{\text{water stress}}$ value was defined as the slope of the upper boundary line relating yield and Δ when N was not limiting (Fig. 1). The upper boundary approach assumes that the line defines the maximum yield for a given amount of available water (Webb, 1972). The negative relationship between Δ‰ and corn yield shown in Fig. 1 is typical for a C_4 plant, whereas for a C_3 plant (such as wheat) the opposite relationship has been observed (Clay et al., 2001a).

To calculate the $\delta\Delta_{\text{N stress}}$ value, N fertilizer should be applied at a range of rates to

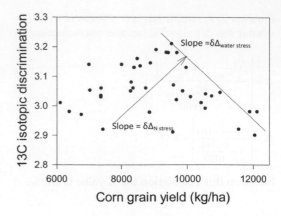

Fig. 1. Relationship between corn yield and Δ in an N rate and watering experiment conducted between 2002 and 2004. Analysis of data from this study is available in Kim et al. (2008).

selected treatments. In these plots, plant populations must be constant. Data where N fertilizer increases yields can be used to calculate the $\delta\Delta_{\text{N stress}}$ value. If a yield response is not observed, then these data should not be included in the calculations. The $\delta\Delta_{\text{N stress}}$ value was the slope of the linear line relating Δ and yield of unfertilized and well-fertilized plants.

In the experiment shown in Fig. 1, the N rates were 0, 50, 150, and 200 kg N ha^{-1}, and the optimum corn yield is the maximum yield observed in the data set. Yield loss due to N and water stress are calculated using a six step approach which includes:

1. Define the upper boundary line based on two points [(11903, 2.97‰) (9250, 3.211‰)] (Fig. 1). $Y = 4.047 - 0.0000905x$; this line defines yield when water stress limits yield (Webb, 1972).

2. Define the N response line in Fig. 1. Two point used in this analysis are (9683, 3.178‰) and (7570, 2.97‰). Based on these points, the N response line was $y = 2.225 + 0.0000984x$.

3. Determine the line with a slope of $\delta\Delta_{\text{N stress}}$ that intersected the measured yield at a given measured Δ value. For example, if yield is 8000 and $\Delta = 3.10$‰, determine the y-intercept. This is determined by solving for the y-intercept in the equation $3.10 = b + 0.0000984(8000)$. When solved, b = 2.313, and the equation was $y = 2.313 + 0.0000984x$.

4. Determine the x and y values at the intersection between the upper boundary line ($y = 4.047-0.00000905x$) and your new N

response line ($y = 2.313 + 0.0000984x$). This was solved by setting the two equations equal to each other ($2.313 + 0.0000984x = 4.047 - 0.00000905x$). When solved x was 9180 kg corn grain ha^{-1}.

5. The yield at the intersection is the maximum yield for a given environment, and therefore YLNS was the difference between the yield at the intersection and the measured yield (YLNS = 1180 kg ha^{-1} = 9180 – 8000).

6. The YLWS is the difference between the yield at the intersection and an optimum yield (YLWS = 2,820 kg ha^{-1} = 12,000 – 9n180).

A similar approach is used in C_3 plants; however, the upper boundary line is defined differently. Differences between the C_4 and C_3 plants are attributed to water stress reducing Δ in C_4 plant and increasing Δ in C_3 plants.

Quantifying Carbon Turnover and Soil Health

Different plant photosynthetic systems have different impacts on the amount of ^{13}C fractionation that is observed in soil. C_4 plants have ^{13}C discrimination values (Δ) that range from 2 to 4‰ and $\delta^{13}C$ values ranging from –12 to –14‰, whereas C_3 plants have ^{13}C discrimination values (Δ) that range from 16 to 18‰ and $\delta^{13}C$ values ranging from –24 to –26‰. The following discussion provides an example for tracking the fate of SOC in soil when the new carbon added to the soil has a different ^{13}C isotopic signature than the old SOC. For example, the initial SOC at the beginning of the experiment is –18‰ and the new carbon has a value of –12‰. These differences have been used to track SOC additions and losses, using Eq. [12 and 13] (Balesdent et al., 1988; Mamani-Pati et al., 2010).

$$F_{C3} = \left(\left| \frac{\delta^{13}C_{\text{soil sample}} - \delta^{13}C_{c4}}{\delta^{13}C_{c4} - \delta^{13}C_{c3}} \right| \right) \quad [12]$$

where $\delta^{13}C_{\text{soil sample}}$ is the $\delta^{13}C$ value of the collected soil sample, $\delta^{13}C_{c4}$ is the $\delta^{13}C$ value of the C_4 plant material, and $\delta^{13}C_{c3}$ is the $\delta^{13}C$ value of the C_3 plant material.

This calculation assumes that ^{13}C isotopic fractionation does not occur during the mineralization of C_4 and C_3 plant materials. A similar equation is used to define the relative

amount of new carbon incorporated into the soil organic C.

$$F_{\text{old carbon}} = \left(\left| \frac{\delta^{13}C_{\text{soil sample}} - \delta^{13}C_{\text{new carbon}}}{\delta^{13}C_{\text{new carbon}} - \delta^{13}C_{\text{initial carbon}}} \right| \right) \quad [13]$$

where $\delta^{13}C_{\text{soil sample}}$ is the $\delta^{13}C$ value of the collected soil sample, $\delta^{13}C_{\text{new carbon}}$ is the $\delta^{13}C$ value of the new plant carbon added to the soil, and $\delta^{13}C_{\text{initial carbon}}$ is the $\delta^{13}C$ value of original SOC.

Many studies have assumed that the $\delta^{13}C_{\text{initial carbon}}$ and the $\delta^{13}C$ of this same carbon (SOC$_{\text{remaining}}$) at some time period in the future are identical. In some experiments this may be true; however, frequently the ^{12}C component of the SOC mineralizes faster than the ^{13}C component of the SOC (called fractionation) (Clay et al., 2007), resulting in a gradual increase in the SOC $\delta^{13}C$ value. This assumption was tested in SOC experiments that including no-plant control areas. This research showed that a failure to account for ^{13}C fractionation resulted in overestimating losses from C_3 and underestimating losses from C_4 plants (Clay et al., 2006, 2007, 2015).

Problem 5. If ^{13}C fractionation does not occur during SOC mineralization, determine the proportion of new carbon remaining in soil if the $\delta^{13}C$ value of a soil sample collected after 1 yr is –16‰, the $\delta^{13}C$ value of the old carbon is –19‰ (soil before the experiment was started), and the $\delta^{13}C$ value of the new carbon is –12‰.

$$F_{\text{old carbon}} = \left(\left| \frac{\delta^{13}C_{\text{soil sample}} - \delta^{13}C_{\text{new carbon}}}{\delta^{13}C_{\text{new carbon}} - \delta^{13}C_{\text{old carbon}}} \right| \right)$$

$$F_{\text{old carbon}} = \left(\left| \frac{-16 - (-12)}{-12 - (-19)} \right| \right) = \frac{4}{7} = 0.57$$

$$F_{\text{new carbon}} = 1 - F_{\text{old carbon}} = 0.43$$

If ^{13}C fractionation in the SOC occurred during the study, then the problem becomes much more difficult. For example, if during the study the old carbon $\delta^{13}C$ value increased from –19‰ to –18.5‰, then the fraction of new carbon incorporated into SOC will decrease from 0.43 to 0.385 {[–16 – (–12)]/[–12 – (–18.5)]}. Clay et al. (2007) evaluated ^{13}C fractionation during SOC mineralization and reported that (i) ^{13}C fractionation during fresh biomass mineralization was not detected, and (ii) not accounting for ^{13}C fractionation during SOC mineralization resulted in (a) overestimating C_4 plant biomass stability, and (b) overestimating SOC mineralization from C_3 plants.

Accounting for fractionation may require accurate measurements of the amount of above- and belowground nonharvested carbon (NHC) that is returned to the soil. Accurate measurements of aboveground biomass can be obtained by harvesting and measuring surface residue from a prescribed area. However, most belowground NHC values are negatively biased because roots and root exudates are difficult to measure. Underestimating belowground biomass can have a profound impact on the calculated budget. For example, Barber (1979) reported that 2.4% of the SOC was mineralized annually and that 37% of the root biomass was converted to SOC. This calculation was based on an assumption that the root-to-shoot ratio was 0.17 (0.17 kg of belowground biomass for each 1 kg of aboveground biomass), which is much lower than Johnson et al. (2006) and Liska et al. (2014). If the root-to-shoot ratio was doubled from 0.17 to 0.34, then the estimated amount of root biomass added to the soil would have been doubled. Under these conditions, the contributions of roots to SOC would have been reduced from 37 to 18.5%. The conversion of 18.5% of the roots to SOC is similar to conversion of surface residues to SOC reported by Clay et al. (2010).

To separate SOC into two pools, $SOC_{retained}$ and plant carbon retained (PCR), Clay et al. (2006, 2007, 2010, 2015) proposed a three-pool conceptual model containing atmospheric CO_2, NHC, and SOC. The advantages of this approach include that (i) C pools can be experimentally measured, (ii) the method does not rely on laboratory-defined SOC turnover rate constants, and (iii) the turnover rates can be easily expanded to other sites. As with Barber (1979), the major flaws with this approach were the reliance on belowground biomass estimates and the long period of time required to generate measureable changes in SOC. Clay et al. (2015) used this basic concept to develop C budgets for a South Dakota field site. This approach is based on the following equations:

$$\delta^{13}C_{soil\ final} = [PCR_{incorp}(\delta^{13}C_{PCR}) + SOC_{retained}(\delta^{13}C_{SOC\ retained})] \quad [14]$$

$$(PCR_{incorp} + SOC_{retained})$$

$$SOC_{final} = PCR_{incorp} + SOC_{retained} \quad [15]$$

$$SOC_{initial} = SOC_{retained} + SOC_{lost} \quad [16]$$

$$SOC_{retained} = [SOC_{final}(\delta^{13}C_{soil\ final} - \delta^{13}C_{PCR})] \quad [17]$$

$$(\delta^{13}C_{SOCretained} - \delta^{13}C_{PCR})$$

$$PCR_{incorp} = SOC_{final}(\delta^{13}C_{soil\ final} - \delta^{13}C_{SOCretained}) \quad [18]$$

$$(\delta^{13}C_{PCR} - \delta^{13}C_{SOCretained})$$

$$\delta^{13}C_{SOC\ retained} = \delta^{13}C_{soil\ initial} + \varepsilon_{SOC} \ln(SOC_{retained}/SOC_{initial}) \quad [19]$$

where, $\delta^{13}C_{soil\ final}$ was the $\delta^{13}C$ value of the bulk SOC when the experiment was completed; PCR_{incorp}, which was not measured, was the plant carbon retained in the soil that was incorporated into SOC; $\delta13C_{PCR}$ was the $\delta^{13}C$ value of the plant material retained in the soil after mineralization; $SOC_{retained}$, which was not measured, was the amount of initial SOC ($SOC_{initial}$) that was not mineralized during the study; and $\delta^{13}C_{SOCretained}$, which was not measured, was the associated $\delta^{13}C$ value. SOC_{final} was SOC at the end of the study.

Equation [17] cannot be solved because it contains two unknowns, $SOC_{retained}$ and $\delta^{13}C_{SOC\ retained}$. To develop additional independent equations, Eq. [18 and 19] were derived. Equation [19] is used to calculate the Rayleigh fractionation constant, E_{SOC}, from no-plant control areas (Balesdent and Mariotti, 1996; Accoe et al., 2002; Fukada et al., 2003; Clay et al., 2007; Spence et al., 2005). Because plants are not growing in these areas, the change in $\delta^{13}C$ is due to isotopic fractionation. For example, if $\delta^{13}C_{SOC\ retained}$ was −16.5‰, $\delta^{13}C_{soil\ initial}$ was −16.9‰, $SOC_{retained}$ was 27 Mg ha^{-1}, and $SOC_{initial}$ was 30 Mg ha^{-1}, then Eq. [19] was used to derive an equation for E_{SOC} [$(\delta^{13}C_{SOC\ retained} - \delta^{13}C_{soil\ initial})/(\ln (SOC_{retained}/SOC_{initial})]$, and calculated the ε_{SOC} value {−3.80=[−16.5 −(16.9)]/ln(27/30)}.

The calculated $SOC_{retained}$, SOC_{lost}, and PCR_{incorp}, values were then determined by using a program that provided an iterative solution to Eq. [16, 17, 18, and 19]. In the iteration, a value was provided for the for $\delta^{13}C_{SOCretained}$, which was then used to calculate $SOC_{retained}$ using Eq. [19], PCR using Eq. [18], and SOC_{lost} using Eq. [17]. In the second iteration, a new guess was made for $\delta^{13}C_{SOCretained}$. This process was repeated until the difference between the calculated $\delta^{13}C_{SOCretained}$ and $SOC_{retained}$ values in two adjacent iterations were minimized. Not all problems can be directly solved. With iterations, when the difference between calculated values in adjacent iterations approaches zero, you have arrived at the solution.

Problem 6. In a plot you are attempting to define C turnover. If $\delta^{13}C_{\text{soil final}}$ is $-15‰$, $\delta^{13}C_{PCR}$ is $-12.5‰$, $\delta^{13}C_{\text{soil initial}}$ is $-15.7‰$, SOC_{final} is 50,000 kg SOC/ha, and SOC_{initial} is 48,000 kg/ha. How much plant carbon (PCR) was incorporated into the soil?

When considering this experiment, it is important to consider that $SOC_{\text{final}} = SOC_{\text{retained}} + PCR$ and that SOC_{retained} by definition is $<SOC_{\text{inital}}$.

In an adjacent experiment where plant growth is prevented, SOC_{initial}, SOC_{retained}, $\delta^{13}C_{\text{soil initial}}$, and $\delta^{13}C_{\text{SOCretained}}$ were 50,000, 45,000, $-16‰$, and $-15.5‰$, respectively.

1. Determine ε_{SOC} using the values for the adjacent experimental plot: $= \delta^{13}C_{\text{soil initial}} + \varepsilon_{SOC} \ln (SOC_{\text{retained}}/SOC_{\text{initial}})$.

$$= \frac{\delta^{13}C_{\text{SOC retained}} - \delta^{13}C_{\text{soil initial}}}{\ln\left(\dfrac{SOC_{\text{retained}}}{SOC_{\text{initlal}}}\right)}$$

$$= -4.75$$

2. Using the ε_{SOC} value you just determined, estimate $\delta^{13}C_{\text{SOC retained}}$ in your plot. Solve, $\delta^{13}C_{\text{SOC retained}} = \delta^{13}C_{\text{soil initial}} + \varepsilon_{SOC} \ln (SOC_{\text{retained}}/SOC_{\text{initial}})$ by guessing that SOC retained is 44,000 kg/ha.

$$-15.7 + (-4.75)\ln\frac{44,000}{48,000}$$

$$= -15.29$$

3. Estimate SOC_{retained} using the values given for your plot. Solve $SOC_{\text{retained}} = [SOC_{\text{final}}(\delta^{13}C_{\text{soil final}} - \delta^{13}C_{PCR})], (\delta^{13}C_{\text{SOC retained}} - \delta^{13}C_{PCR})$.

$$= \frac{50,000\left[-15 - (-12.5)\right]}{\left[-15.29 - (-12.5)\right]}$$

$$= 44,800 \text{ kg SOC/ha}$$

Iteration 2.

4. Repeat Step 2.

$$-15.7 + (-4.75)\ln\frac{44,800}{48,000}$$

$$= -15.37$$

5. Repeat Step 3.

$$= \frac{50,000\left[-15 - (-12.5)\right]}{\left[-15.37 - (-12.5)\right]}$$

$$= 43,554 \text{ kg SOC/ha}$$

Iteration 3.

6. Estimate SOC_{retained}.

$$= (43,554 + 44,800)/2$$

$$= 44,177$$

7. Repeat Step 4.

$$-15.7 + (-4.75)\ln\frac{44,177}{48,000}$$

$$= -15.31$$

8. Repeat Step 5.

$$= \frac{50,000\left[-15 - (-12.5)\right]}{\left[-15.31 - (-12.5)\right]}$$

$$= 44,483 \text{ kg SOC/ha}$$

Iteration 4:

9. Estimate SOC_{retained}.

$$= (44,177 + 44,483)/2$$

$$= 44,330$$

10. Repeat Step 7.

$$-15.7 + (-4.75)\ln\frac{44,330}{48,000}$$

$$= -15.32$$

-15.32 is close to previous calculation (from Step 7), so use this in your final calculation.

11. Repeat Step 8.

$$= \frac{50,000\left[-15 - (-12.5)\right]}{\left[-15.32 - (-12.5)\right]}$$

$$= 44,326 \text{ kg SOC/ha}$$

12. Determine Plant carbon incorporated (PCR).

$SOC_{\text{final}} = PCR_{\text{incorp}} + SOC_{\text{retained}}$

$PCR = SOC_{\text{final}} - SOC_{\text{retained}}$

$PCR = 50,000 - 44,326 \text{ kg/ha} = 5673 \text{ kg/ha}$

Based on this analysis, 5673 kg/ha of new carbon has been integrated into the soil, and 44,326 kg SOC/ha that was in the soil at the beginning of the experiment (48,000 kg/ha) remains in the soil at the completion of the experiment.

Determining Fertilizer with Nitrogen-15 Enriched Fertilizer

The use of materials enriched with ^{15}N and ^{13}C can be used to overcome some of limitations associated with the natural abundance approach. The primary strength of enriched materials is that the labeled signal can be followed for multiple years. The disadvantages include: (i) high cost, (ii) chemical and biological fractionation is not considered, (iii) these materials are not applied with commercial equipment, and (iv) the experiments are generally conducted in small areas. When using enriched materials, contamination becomes a larger problem and care must be used to minimize contamination. Labeled materials can be used to track the fate of targeted materials. Many of the calculation techniques of this approach are outlined in Cabrera and Kissel (1989), Knowles and Blackburn. (1993), Barraclough (1995), and Ryabenko (2013).

Problem 7. Develop a 5 At%^{15}N solution from 10 At%^{15}N fertilizer.

1. Determine the molecular weight of the labeled fertilizer.

$$= (0.10)\left(\frac{15 \text{ g}}{\text{mole}}\right) + (0.90)\left(\frac{14 \text{ g}}{\text{mole}}\right) = \frac{14.1 \text{ g}}{\text{mole}}$$

2. Determine the amount of labeled fertilizer to add:

$$5 \text{At}\% = \frac{x}{100} 10 \text{ At}\% + \left(\frac{100-x}{100}\right) 0.366$$

$x = 48.10$; moles of $x = 0.4810$

    moles of $^{14}N = 1 - 0.4810 = 0.5180$

3. Determine amount of labeled and unlabeled fertilizer.

$$\left(0.4810 \text{ moles of } ^{15}N\right)\left(\frac{14.10 \text{ g}}{\text{mole}}\right) =$$

    6.782 g labeled fert.

$$\left(51.90 \text{ moles unlabeled}\right)\left(\frac{14.0067 \text{ g}}{\text{mole}}\right) =$$

    7.2695 g unlabeled fert.

Problem 8. Corn grain has a ^{15}N label of 2At%^{15}N. If the fertilizer had a label of 5At%, how much N was derived from the labeled fertilizer? The corn grain yield was 10,000 kg ha^{-1}, the N content of the plant was 1.5%, and N was applied at the rate of 100 kg/ha. Use this same data for Problems 9, 10, and 11 below.

1. Determine the molecular weight of the N in the plant.

$$= 0.02\left(\frac{15 \text{ g}}{\text{mole}}\right) + 0.98\left(\frac{14 \text{ g}}{\text{mole}}\right) = 14.02 \text{ g/mole}$$

2. Determine total N taken up by the plant.

$$= \left(\frac{0.015 \text{ g N}}{\text{g grain}}\right)\left(\frac{10,000}{\text{ha}}\right) = \frac{150 \text{ kg N}}{\text{ha}}$$

3. Determine the grams of ^{15}N in the grain.

$$= \frac{150 \text{ kg N}}{\text{ha}} \frac{1000 \text{ g}}{1 \text{k g}} \frac{\text{mole N}}{14.02 \text{ g}} \frac{0.02 \text{ moles}^{15}N}{1 \text{ mole N}}$$

$$\frac{15 \text{g}}{1 \text{ mole}^{15}N} = \frac{3209.7 \text{ g}^{15}N}{\text{ha}}$$

4. Determine the total amount of ^{15}N applied in the fertilizer.

$$\left(\frac{100 \text{ kg N}}{\text{ha}}\right)\left(\frac{1000 \text{ g}}{1 \text{ kg}}\right)\left(\frac{\text{mole}}{14.02 \text{g N}}\right)\left(\frac{0.05 \text{ mole}^{15}N}{1 \text{mole fert}}\right)$$

$$\left(\frac{15 \text{g}}{1 \text{ mole}^{15}N}\right) = 5349.5 \text{ g }^{15}N$$

5. What is the percentage of ^{15}N applied in the fertilizer taken up by the plant?

$$= 100\% \frac{3207.7 \text{ g/ha}}{5349.35 \text{ g/ha}} = 59\%$$

Based on this calculation, you may assume that the fertilizer use efficiency was 59%. However, this would be incorrect. This calculation does not consider the ^{15}N derived from the soil. To account for soil-derived N, the experiment needs a 0 N rate, where similar calculations are made.

Problem 9. Determine the amount ^{15}N in an unfertilized control plot. In this experiment the At% is 0.37, yield is 8000 kg/ha, and the N percent in the corn plant is 1.25%.

1. Determine the molecular weight of the N in the plant.

$$= 0.0037\left(\frac{15\text{ g}}{\text{mole}}\right) + 0.9963\left(\frac{14\text{ g}}{\text{mole}}\right)$$

$$= 14.0037\text{ g/mole}$$

2. Determine the total N taken up by the plant.

$$= \left(\frac{0.0125\text{ g N}}{\text{g grain}}\right)\left(\frac{8000}{\text{ha}}\right) = \frac{100\ \text{kg N}}{\text{ha}}$$

3. Determine the grams of ^{15}N in the grain that was derived from the soil.

$$= \frac{100\text{ kg N}}{\text{ha}}\ \frac{1000\text{ g}}{1\text{kg}}\ \frac{\text{mole N}}{14.0037\text{ g}}\ \frac{0.0037\text{ moles}^{15}\text{N}}{1\text{ mole N}}$$

$$\frac{15\text{ g}}{1\text{ mole }^{15}\text{N}} = \frac{396.3\text{ g}^{15}\text{N}}{\text{ha}}$$

4. Based on results from Problem 8, calculate ^{15}N in the plant derived from the ferilizer.

Total N = N from fertilizer + N from soil
N from fertilizer = 3209.7 − 396.3 = 2813.4

5. What is the percentage of fertilizer N taken up by the plant?

$$= 100\%\frac{2813.4\text{ g/ha}}{5349.35\text{ g/ha}} = 52\%$$

Problem 10. Determine the percent of N from the plant using the equation: % fertilizer = [(N_u − N_t)/(N_u − $N_f n$)]100, where N_u is the At% of plants that were not fertilized, N_t is the At% of the plant in the fertilized treatment, N_f is the At% in the fertilizer, and n is the discrimination factor (generally assumed to be 1, no discrimination).

Based on the above, N_u = 0.37 At%, N_t = 2 At%, and Nf = 5 At%:

$$\%\text{ fertilizer} = \left(\frac{N_u - N_t}{N_u - N_f n}\right)100$$

$$\%\text{ fertilizer} = \left(\frac{0.37 - 2}{0.37 - 5}\right)100 = 35\%$$

where N_u is the ^{15}N At% of the unfertilized plant, N_t is the At% of the unfertilized plant, Nf is the At% of the fertilizer, and n is ^{15}N discrimination.

The amount of fertilizer N in the plant is then determined by multiplying the ratio by the amount of N in the plant. Because the amount of N in the plant is 150 kg N/ha (0.015 × 10,000 kg/ha), the amount of grain N derived from the fertilizer was 52.5 kg N/ha (0.35 × 150/ha). The N was applied at a rate of 100 kg/ha which results in an efficiency of 52% (100 × 52.5/100). This value is very similar to the fertilizer efficiency

value reported in Problem 9. Addition information on the use of enriched isotopic techniques is available in Hauck and Bremner (1976).

Determine Fertilizer Efficiency with Nitrogen-15 Depleted Fertilizer

The depleted N method is very similar to the enriched approach; however, the fertilizer is depleted in ^{15}N. In this method, the ^{15}N At% values generally range from 0.003 to 0.01. The amount of plant N derived from the fertilizer is calculated with the equation,

$$\%\text{ fertilizer} = \left(\frac{N_u - N_t}{N_u - N_f n}\right)100 \tag{20}$$

where N_u is the ^{15}N At% of the unfertilized plant, N_t is the At% of the unfertilized plant, N_f is the At% of the fertilizer, and n is ^{15}N discrimination. This is the identical equation that was used in Problem 10. One of the primary disadvantages with the depleted approach is the difficulty of tracking N from one season to the next (Clay and Malzer, 1993). The N use efficiency values using depleted, enriched, and the nonisotopic difference method are often highly correlated to each other. Additional information on this approach and comparisons with the nonisotopic difference approach are available in Harmsen (2003a, 2003b).

Problem 11. Determine the efficiency by nonisotopic difference techniques.

N use efficiency =

$$\left(\frac{\text{N uptake}_{\text{fertilized}} - \text{N uptake}_{\text{0 N control}}}{\text{N applied}}\right)$$

$$= \left(\frac{150 - 100}{100}\right) = 50\%$$

Summary

In summary, a critical component of feeding a growing world population is the development of new tools that can be used to assess management impacts on soil and plant processes. The use of stable isotopic approaches should be put into this toolbox. This chapter discusses the use of the enriched N fertilizers and natural abundance techniques for assessing yield responses to multiple stresses as well as determining SOC turnover. The importance of soil organic C in soil fertilizer recommendations has been noted in fertilizer recommendations. Soil organic C is directly related to the soil's water holding capacity and yield potential (Clay et al., 2014). The largest problem with the use of ^{15}N enriched fertilizers is cost, which results in plots that are very small. The size of the plot limits the technique's usefulness, because extensive soil sampling in a small area can result in compaction and reduced yields. One approach that can be used to overcome problems using the enriched fertilizer approach is the natural abundance approach. However, in this technique the differences between treated and nontreated areas are very small. Because the natural abundance approach does not require the purchase of enriched or depleted fertilizer, commercially available equipment can be used and the plot sizes can be very large. In addition, this approach can be integrated into on-farm research projects.

References

Accoe, F., P. Boeckx, O. Van-Cleemput, G. Hofman, Y. Zhang, R. Li, and C. Guanxiong. 2002. Evolution of the delta ^{13}C signature related to total carbon content and carbon decomposition rate constants in a soil profile under grassland. Rapid Commun. Mass Spectrom. 16:2184–2189. doi:10.1002/rcm.767

Balesdent, J., and A. Mariotti. 1996. Measurement of soil organic matter turnover using ^{13}C natural abundance. In: T.W. Boutton and S. Yamasaki, editors, Mass spectrometry of soils. Marcel Dekker, New York. p. 83–112.

Balesdent, J., G.H., Wagner, and A. Mariotti. 1988. Soil organic matter turnover in long-term experiments as revealed by carbon-13 natural abundance. Soil Sci. Soc. Am. J. 52:118–124. doi:10.2136/sssaj1988.03615995005200010021x

Barber, S.A. 1979. Corn residue management and soil organic matter. Agron. J. 71:625–627. doi:10.2134/agronj1979.00021962007100040025x

Barraclough, D. 1995. [15]N isotopic dilution techniques to study soil nitrogen transformation and plant uptake. Fert. Res. 42:185–192.

Cabrera, M.L., and D.E. Kissel. 1989. Review and simplifications of calculations in ^{15}N tracer studies. Fert. Res. 20:11–15. doi:10.1007/BF01055396

Carelli, M.L.C., J.I. Fahl, P.C.O. Trivelin, and R.B. Queiroz-Voltan. 1999. Carbon isotope discrimination and gas exchange in *Coffea* species grown under different irradiance regimes. Rev. Bras. Fisiol. Vegetal 11:63–68.

Clapp, C.E., R.A. Allmaras, M.F. Layese, D.R. Linden, and R.H. Dowdy. 2000. Soil organic carbon and 13-C abundance as related to tillage, residue, and nitrogen fertilizer under continuous corn

management in Minnesota. Soil Tillage Res. 55:127–142. doi:10.1016/S0167-1987(00)00110-0

Clay, D.E., C.G. Carlson, S.A. Clay, V. Owens, T.E. Schumacher, and F. Mamani-Pati. 2010. Biomass estimation approach impacts on calculated SOC maintenance requirements and associated mineralization rate constants. J. Environ. Qual. 39:784–790. doi:10.2134/jeq2009.0321

Clay, D.E., C.G. Carlson, S.A. Clay, J. Chang, and D.D. Malo. 2005. Soil organic C maintenance in a corn (*Zea mays* L.) and soybean (*Glycine max* L.) as influenced by elevation zone. J. Soil Water Conserv. 60:342–348.

Clay, D.E., C.G. Carlson, S.A. Clay, C. Reese, Z. Liu, and M.M. Ellsbury. 2006. Theoretical derivation of new stable and non-isotopic approaches for assessing soil organic C turnover. Agron. J. 98:443–450. doi:10.2134/agronj2005.0066

Clay, D.E., C.E. Clapp, C. Reese, Z. Liu, C.G. Carlson, H. Woodard, and A. Bly. 2007. [13]C fractionation of relic soil organic C during mineralization effects calculated half-lives. Soil Sci. Soc. Am. J. 71:1003–1009. doi:10.2136/sssaj2006.0193

Clay, S.A., D.E. Clay, D. Horvath, J. Pullis, C.G. Carlson, S. Hansen, and G. Reicks. 2009. Corn (*Zea mays*) responses to competition: Growth alteration vs. limiting factor. Agron. J. 101:1522–1529. doi:10.2134/agronj2008.0213x

Clay, D.E., S.A. Clay, J. Jackson, K. Dalsted, C. Reese, Z. Liu, D.D. Malo, and C.G. Carlson. 2003. Carbon-13 discrimination can be used to evaluate soybean yield variability. Agron. J. 95:430–435. doi:10.2134/agronj2003.0430

Clay, D.E., S.A. Clay, D.J. Lyon, and J.M. Blumenthal. 2005. Can ^{13}C discrimination in corn (*Zea mays*) grain be used to characterize intra-plant competition for water and nitrogen? Weed Sci. 53:23–29.

Clay, D.E., S.A. Clay, Z. Lui, and C. Reese. 2001b. Spatial variability of Carbon-13 isotopic discrimination in corn (*Zea mays*). Commun. Soil Sci. Plant Anal. 32:1813–1828. doi:10.1081/CSS-120000252

Clay, D.E., D.S. Clay, K. Reitsma, B. Dunn, A. Smart, G. Carlson, D. Horvath, and J. Stone. 2014. Does the conversion of grasslands to row crop production in semi-arid areas threaten global food security? Global Food Secur. 3:22–30.

Clay, D.E., R.E. Engel, D.S. Long, and Z. Liu. 2001a. Using C13 discrimination to characterize N and water responses in spring wheat. Soil Sci. Soc. Am. J. 65:1823–1828. doi:10.2136/sssaj2001.1823

Clay, D.E., and G.L. Malzer. 1993. Comparison of two chemical methods for extracting residual N fertilizer. Biol. Fertil. Soils 15:179–184. doi:10.1007/BF00361608

Clay, D.E., G. Reicks, C.G. Carlson, J. Miller, J.J. Stone, and S.A. Clay. 2015. Residue harvesting and yield zone impacts C storage in a continuous corn rotation. J. Enron. Qual. 44:803–809.

Fox, R.H., and W.P. Piskielek. 1995. The relationship between corn grain yield goal and economic optimum nitrogen fertilizer rates. Agron. Ser. 136, Pennsylvania State Univ. University Park, PA.

Fukada, T.K.M., K.M. Hiscock, P.F. Dennis, and T. Grischek. 2003. A dual isotope approach to identify denitrification at a river-band infiltration site. Water Res. 37:3070–3078. doi:10.1016/S0043-1354(03)00176-3

Harmsen, K. 2003a. A comparison of the isotopic dilution and the difference for estimating fertilizer nitrogen recovery fractions in crops: I. Plant uptake and loss of nitrogen. Wageningen J. Life Sci. 50:321–347. doi:10.1016/S1573-5214(03)80015-5

Harmsen, K. 2003b. A comparison of the isotopic dilution and the difference for estimating fertilizer nitrogen recovery fractions in crops: II. Mineralization and immobilization of nitrogen. Wageningen J. Life Sci. 50:349–381. doi:10.1016/S1573-5214(03)80016-7

Harding, W.R., and R.C. Hart. 2011. Use of stable isotope analysis to describe aquatic foodwebs in the Kruger National Park. Report to the Water Research Committee by DH Environmental Consulting. Summerset West, South Africa. http://www.wrc.org.za/Documents/KV%20256.pdf (accessed 21 Apr. 2015).

Hauck, R.D., and J.M. Bremner. 1976. Use of tracers for soil and fertilizer nitrogen research. Adv. Agron. 28:219–266. doi:10.1016/S0065-2113(08)60556-8

Hay, J.M. 2004. An introduction to isotopic calculations. Wood Hole Oceanographic Institute, Wood Hole MA. http://www.whoi.edu/cms/files/jhayes/2005/9/IsoCalcs30Sept04_5183.pdf (accessed 29 Oct. 2015).

Henaux, V., L.A. Powell, M.P. Vrtiska, and K.A. Hobson. 2012. Establishing winter origins of migrating lesser Snow Geese using stable isotopes. Avian Conserv. Ecol. 7(1):5. doi:10.5751/ACE-00515-070105

International Atomic Energy Agency. 2013. Applications of Isotopic Techniques for Nutrient Dynamics in River Basins, IAEA-TECHDOC-1695. http://www-pub.iaea.org/MTCD/Publications/PDF/TE-1695_web.pdf (accessed 29 Oct. 2015).

Jahren, A.R., and R.A. Kraft. 2008. Carbon and nitrogen isotopes in fast foods: Signatures of corn confinement. Proc. Natl. Acad. Sci. U.S.A 105:17855–17860. doi:10.1073/pnas.0809870105

Jefferies, R.A., and D.K.L. Mackerron. 1997. Carbon isotope discrimination in irrigated and droughted potato (Solanum tuberosum L.). Plant Cell Environ. 20:124–130. doi:10.1046/j.1365-3040.1997.d01-5.x

Johnson, J.M.F., R.R. Allmaras, and D.C. Reicosky. 2006. Estimating source carbon from crop residues, roots, and rhizodeposits using the national grain-yield data-base. Agron. J. 98:622–636. doi:10.2134/agronj2005.0179

Kendal, C. 1989. Tracing N source and cycling in catchments. In: C. Kendal and J.J. McDonnel, editors, Isotope tracers in catchment hydrology. Elsevier, Amsterdam, The Netherlands. p. 519–576.

Kharel, T.P., D.E. Clay, D. Beck, C. Reese, C.G. Carlson, and H. Park. 2011. Nitrogen and water rate effect on winter wheat yield, N and water use efficiency and dough quality. Agron. J. 103:1389–1396. doi:10.2134/agronj2011.0011

Kim, K.-I., D. Clay, S. Clay, G.C. Carlson, and T. Trooien. 2013. Testing corn (Zea Mays L.) preseason regional nitrogen recommendation models in South Dakota. Agron. J. 105:1619–1625. doi:10.2134/agronj2013.0166

Kim, K.-I., D.E. Clay, C.G. Carlson, S.A. Clay, and T. Trooien. 2008. Do synergistic relationships between nitrogen and water influence the ability of corn to use nitrogen derived from fertilizer and soil? Agron. J. 100:551–556. doi:10.2134/agronj2007.0064

Knowles, R. and T.H. Blackburn. 1993. Nitrogen isotope techniques. Academic Press, San Diego, CA.

Liska, A.J., H. Yang, M. Milner, S. Goddars, H. Blanco-Cangui, M.P. Pelton, X.X. Fang, H. Zhu, and A.E. Suyker. 2014. Biofuel from crop residue can reduce soil carbon and increase CO_2 emissions. Nat. Clim. Change 4:398–401. doi:10.1038/nclimate2187

Lory, J.A., and P.C. Scharf. 2003. Yield goal versus delta yield for prediction fertilizer nitrogen needs in corn. Agron. J. 95:994–999. doi:10.2134/agronj2003.0994

Mamani-Pati, F.M., D.E. Clay, C.G. Carlson, and S.A. Clay. 2010. Calculating soil organic carbon maintenance using stable and isotopic approaches: A review. In: E. Lichtfouse, editor, Sociology, organic farming, climate change and soil science. Springer, Amsterdam, The Netherlands, p. 189–216.

Park, H., D.E. Clay, R.G. Hall, J.S. Rohila, T.P. Kharel, S.A. Clay, and S. Lee. 2014. Winter wheat quality responses to water, environment, and nitrogen fertilization. Commun. Soil Sci. Plant Anal. 45:1894–1905. doi:10.1080/00103624.2014.909833

Pate, J., and D. Arthur. 1998. ^{13}C analysis of phloem sap carbon: Novel means for evaluating seasonal water stress and interpreting carbon isotope signatures of foliage and trunk wood of Eucalyptus globulus. Oecologia 117:301–311. doi:10.1007/s004420050663

Reese, C., D.E. Clay, S.A. Clay, A. Bich, A. Kennedy, S. Hansen, and J. Miller. 2014. Wintercover crops impact on corn production in semi-arid regions. Agron. J. 106:1479–1488. doi:10.2134/agronj13.0540

Ryabenko. E. 2013. Stable isotopic methods for the study of the nitrogen cycle INTECH. http://cdn.intechopen.com/pdfs-wm/45198.pdf (accessed 28 Oct. 2015).

Smeltekop, H., D.E. Clay., and S.A. Clay. 2002. The impact of sava snail medic cover crop on corn production, stable isotope discrimination, and soil quality. Agron. J. 94:917–924. doi:10.2134/agronj2002.9170

Spence, M.J., S.H. Bottrell, S.I. Thornton, H.H. Richnow, and K.H. Spence. 2005. Hydrochemical and isotopic effects associated with petroleum fuel biodegradation pathways in a chalk aquifer. J. Contam. Hydrol. 79:67–88. doi:10.1016/j.jconhyd.2005.06.003

Tiunov, A.V. 2007. Stable isotopes on carbon and nitrogen in soil ecological studies. Ecology 34:475–489.

Webb, R.A. 1972. Use of the boundary line in the analysis of biological data. J. Hortic. Sci. 47:309–319.

- Subsurface Tile Drainage: Advantages and Its Implications on Environment
- Factors Influencing Subsurface Nutrient Loss
- Management Practices to Reduce Nutrient Loss under Tile-Drained Conditions

Soil and Fertilizer Management Practices to Control Nutrient Losses under Subsurface Tile-Drained Conditions

Amitava Chatterjee

Subsurface tile drainage is the common water management tool in the northern Great Plains. Although installing tile drainage facilitates crop production by removing excess water, nevertheless tile drainage has the potential to increase nutrient loss. Rainfall distribution and crop and soil management practices play a role in nutrient movement through the soil profile. Precipitation amount, intensity, and temporal distribution control drainage volume and concentration in discharged water. Adoption of a 4R nutrient stewardship program, cover crop, and conservation tillage practices have the potential to increase soil nutrient retention.

Subsurface tile drainage is the most common and extensive agricultural water management practices observed in the US Midwest, Canada, and northern Europe to remove excess water that is seasonally or perennially wet (Pavelis, 1987). Agricultural drainage made it possible to transform large swamp lands of Ohio, Indiana, Illinois, and Ohio to very productive agricultural land (Pavelis, 1987). Crops need both air and water in the root zones, but excess water filled out all the soil pore space and restricted aeration. Subsurface tile drain removes standing or excess water of agricultural land with poor internal drainage and a high water table. Under saline conditions, subsurface drainage is required to prevent the water table from rising too near to the soil and to remove salts (Schwab et al., 1996).

Wetness has a strong influence on soil strength; particularly wet clay soil becomes very weak to support traffic. In a temperate climate, soil warms up slowly in the spring and seed germination suffers (Troeh et al., 2004). Poor aeration creates anoxic soil environment and a low oxygen diffusion rate. Loss of nitrogen (denitrification) peaks in this anoxic environment, particularly under

---

Abbreviations: ET, evapotranspiration; FWMC, flow-weighted mean concentration; SOM, soil organic matter.

Amitava Chatterjee, Dep. of Soil Science, Walster 133, 1402 Albrecht Blvd., North Dakota State Univ., Fargo, ND 58108 (amitava.chatterjee@ndsu.edu).

doi:10.2134/soilfertility.2014.0009

Soil Fertility Management in Agroecosystems
Amitava Chatterjee and David Clay, Editors

soils with high organic matter content. Poor aeration also reduces the potassium and iron absorption by plant roots. Wetness also favors diseases like root rot.

Subsurface tile drainage offers an opportunity to manage the water table under poorly drained conditions. However, subsurface drainage also invites the risk of nutrient loss with improper crop, soil, and fertilizer management practices. It is important to understand the circumstances of increase nutrient losses under tile drainage and how different management practices influence nutrient movement under tile-drained condition.

Subsurface Tile Drainage: Advantages and Its Implications on Environment

Many soils in the upper Midwest, as well as soils in other regions of the United States, have inherent drainage issues and suffer from water-logged conditions (Busman and Sands, 2002). According to the 2012 Census of Agriculture, the number of US farms with tile drainage systems was 217,931 (14% of total), covering 48.6 million acres (12.5% of total) of cropland. Subsurface drainage has multiple benefits, including the following: (i) a decrease in the length of time of saturated conditions, (ii) an increase in the length of the growing season by removing excess water, (iii) an improvement in trafficability for timely planting and harvesting, (iv) a decrease in surface runoff, (v) an increase in soil aeration, and, finally, (vi) a promotion of rapid root growth (Sands, 2001; Strock et al., 2011; King et al., 2014).

Subsurface drainage does not influence plant available water and removes only "drainable water" that is held in between the saturated and field capacity level. Figure 1 illustrates the sharing of pore space occupied by air and water before and (right) and after (left) draining of soils. Before drainage, there is only a small volume of air-filled pore space with a shallow water table. After draining, the soil has additional pore volume available for infiltration following the curve just after drainage ceases (drainable porosity). More air-filled volume after draining results in less runoff and more infiltration during the next rainfall and ground water recharge than undrained soil. Over time, the curve shifted to the left because of crop water uptake. The drop in water table depth before and after draining is indicated

by the arrow. The extent of water table drop depends on soil structure and texture. For example, sandy soils will have a smaller drop in the water table compared with soils with high clay content. Tile-drained soils have more pore space available for water infiltration and less surface runoff.

Subsurface drainage can increase nitrate (NO_3^-) leaching losses of nutrients below the crop root zone and potentially to groundwater. Water from tile drainage can be discharged into surface ditches or streams. It is believed that NO_3^- in drainage water is a contributor to hypoxic zones in the Gulf of Mexico (Fig. 2). Within the Great Lakes and Corn Belt states, areas of drained cropland were ranked in the following order: first, Illinois; second, Indiana; third, Iowa; fourth, Ohio; seventh, Minnesota; 11th, Michigan; 13th, Missouri; and 16th, Wisconsin (Fausey et al., 1995). Goolsby et al. (1999) estimated that annual total nitrogen (N) and phosphorus (P) exports from the Mississippi River and its tributaries to the Gulf of Mexico is approximately 1.6 and 0.14 million metric tons, respectively. In Minnesota, tile drainage contributes significant N in surface waters (Minnesota Pollution Control Agency, 2013). They reported that during an average precipitation year, row crops under tile-drained fields contribute 67% of the N load in the heavily tiled Minnesota River Basin. According to the Safe Drinking Water Act of 1975 established by the USEPA,

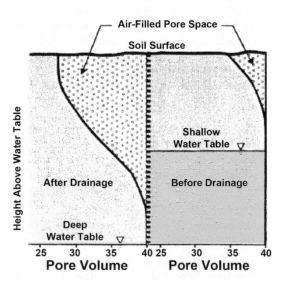

Fig. 1. Differences in air- and water-filled pore space after (left) and before (right) drainage of soil. Adapted from Sands (2001).

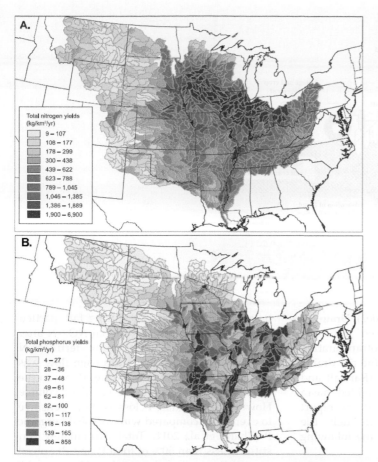

Fig. 2. Total nitrogen (a) and total phosphorus (b) yields (kg/km²/yr) for watersheds in the Mississippi River Basin. From Robertson et al. (2014).

Total nitrogen yields
(kg/km²/yr)

- 9 – 107
- 108 – 177
- 178 – 299
- 300 – 438
- 439 – 622
- 623 – 788
- 789 – 1,045
- 1,046 – 1,385
- 1,386 – 1,889
- 1,900 – 6,900

Total phosphorus yields
(kg/km²/yr)

- 4 – 27
- 28 – 36
- 37 – 48
- 49 – 61
- 62 – 81
- 82 – 100
- 101 – 117
- 118 – 138
- 139 – 165
- 166 – 858

maximum contaminant level of nitrate (NO_3^-) in drinking water is 45 mg/L or 10 ppm NO_3^--N. In Minnesotan rivers and streams between 2000 and 2010, 27% of the 728 samples exceeded the USEPA drinking water standard. Loss of P through tile drainage was initially underestimated compared with surface runoff because of its slow movement and adsorption; however, a review work by Sims et al. (1998) reported significant P leaching and export in deep sandy soils that were tiled. Elevated concentrations of nutrients have the possible consequences of contamination of downstream water resources, such as reservoirs leading to eutrophication of downstream surface waters with possible threats to aquatic ecosystems. These effects are further aggravated in the case of watershed basins receiving high rainfall, enriched in soil organic matter and dominated by corn and soybean (Burkart and James, 1999; Randall and Goss, 2008).

Factors Influencing Subsurface Nutrient Loss

Nutrient losses from a tile-drained field are functions of precipitation, crop and soil characteristics, and tile depth and spacing (Fig. 3). It is important to understand the control of these factors on volume and the concentrations of nutrient transport to understand and develop the sustainable nutrient stewardship program and reduce the nutrient loading into surface water.

Precipitation

Nutrient transport through drainage is a function of (i) the amount of annual or growing season precipitation, (ii) annual temporal distribution of precipitation, and (iii) intensity of an individual rainfall event (Randall and Goss, 2008). Precipitation drives drain flow, and higher drain flow was observed in wet years and lower drain flow in dry years (Randall and Irgavarapu, 1995). In the upper US Midwest, tiles generally flow from mid to late March to mid-July and again in late September through

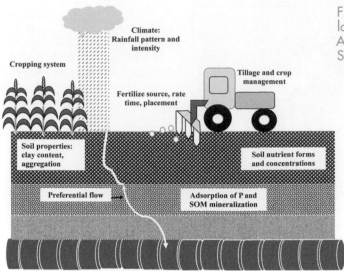

Fig. 3. Factors controlling nutrient losses to subsurface tile drainage. Adapted from King et al. (2014). SOM, soil organic matter.

October (Fig. 4) (Dinnes et al., 2002). Drainage volume is significantly increased during a rainfall in the spring when evapotranspiration losses are small and soil above the tile drains is likely near the field capacity (Randall and Goss, 2008). The term "flow-weighted mean concentration" (FWMC) is used to report average nutrient discharged through tile and can be calculated by using the following equation:

$$FWMC = \frac{\sum_1^n (C_i t_i q_i)}{\sum_1^n (t_i q_i)}$$

where C_i is concentration of the ith sample, t_i is time window for the ith sample, and q_i is flow in the ith sample. Flow-weighted NO_3^- concentrations showed a sharp increase in wet years after a dry spell and can be more than double than those found during an average precipitation year (Randall and Irgavarapu, 1995). Within a year, NO_3^- concentrations in tile drainage can

be significantly increased during fall, particularly in the case of high residual NO_3^--N after harvest and fall application of fertilizer-N. Concentration of P in tile drainage flow was the dominant factor contributing P loss to tile drainage rather than the event flow volume. However, particulate-P loss is more sensitive to event flow compared with dissolved organic P (Zhang et al., 2014). Total particulate-P constituted less than 50% of total P, and dissolved organic P contributed more than 40% of total P concentrations (Beauchemin et al., 1998).

Soil Characteristics

Export of soil nutrients through subsurface drainage begins with downward movement, either by leaching through the soil profile and/or preferential flow through macropores, root channels, or earthworm burrows (Fig. 3). Further, chemical reactions like adsorption and mineralization of nutrients influence nutrient concentrations.

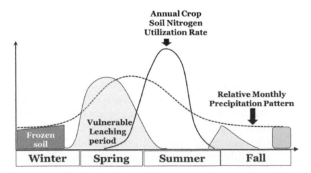

Fig. 4. Vulnerable leaching during different times of the year. Adapted from Dinnes et al. (2002).

On poorly drained and high clay soils, soil moisture controls the NO_3^- production through soil organic matter mineralization. During above normal precipitation, residual NO_3^- concentration can accumulate because of poor plant growth (Randall and Irgavarapu, 1995). The amount of NO_3^- leaching is directly proportional to the rate of soil organic matter mineralization, which is dependent on tillage, residual N after crop harvest, volume of drainage, N uptake by current crop, and previous crop type (legume or nonlegume) (Randall and Goss, 2008).

Soil water can move through the soil profile by preferential/macropore flow and matrix flow, simultaneously. In preferential flow, soil water moves through cracks particularly dominant under unsaturated condition, whereas matrix flow is a more uniform water movement and often can be seen in sandy textured soil (Gburek et al., 2005). Water transport through matrix is slower than preferential flow, but nutrient moving within matrix gets opportunities to be adsorbed by the soil. Extent of preferential flow depends on soil type, soil biota, root system, and tillage. Swell-shrink clay soils develops wide cracks extending from the surface to deep into the soil profile. Earthworms and other burrowing animals develop tunnels, particularly over tile line compared with in between tile lines (Cooley et al., 2013). Root systems of some crops like corn can extend up to 1–3 m (4–6 ft) into the soil, and the decay of these roots can create channels for preferential flow. Compared with conventional tillage practices, long-term no-till favors development of preferential flow paths. Instant increase in leachate P (mostly organic forms) concentrations just after precipitation indicates that preferential flow paths are a significant pathway for P transport, but the total amount of particulate-P can be increased if matrix flow is dominant (King et al., 2014).

Soil texture, organic matter, and other soil constituents (oxides of Fe and Al and carbonates) also control the leaching and retention of nutrients in soil profile. Leaching of P can be increased in organic or coarse-textured mineral soils because of the low concentrations of Fe and Al responsible for P retention (Sims et al., 1998). Accumulation of water on the soil surface diminishes the formation of Fe- and Al-P complexes, leading to a reduction in P binding capacity and increasing the risk of P loss (Zhang et al., 2014). Although, fine-textured soils have greater P sorption capacity than organic and sandy soil (because of low anion exchange capacity), preferential water flow in fine-textured soils can increase the subsurface P losses (King et al., 2014).

Crop and Fertilizer Management Practices

Row crops, particularly corn and soybean, generate higher concentrations of remaining NO_3^- compared with perennial crops that have greater season-long root activity (Randall and Goss, 2008). Model simulation results suggested that a decrease in NO_3^- leaching is possible with diversified rotations (corn–soybean–wheat and corn–rye–soybean–rye) relative to conventional corn–soybean rotation (Tonitto et al., 2007).

Organic manures contribute greater concentration of P losses than chemical fertilizer (Nayak et al., 2009). Mass of NO_3–N or dissolved P lost in subsurface tile drainage increases as application rate increases (Randall and Goss, 2008; King et al., 2014). In addition, nutrient losses can be enhanced by the application of manure, deep sandy soils, high residual nutrients after harvest, or long-term overfertilization (Sims et al., 1998).

Tile Design: Depth, Spacing, and Control Structure

Tile drain spacing and depths are based on soil type, soil permeability, cropping system, desired drainage coefficient (amount of water removal per day), and degree of surface drainage (Wright and Sands, 2001). Changes in average depth and spacings with soil textural class are presented in Table 1 (Schwab et al., 1993). Closer drain spacing results in a higher drainage coefficient and faster drainage but also involves increased cost. Reducing NO_3^- concentrations in tile drains involves reducing the volume of drainage water and increasing denitrification to N_2 gas within the saturated zone of the soil profile by placing the drain at shallower depths (Randall and Goss 2008). Nangia et al. (2010) found a range of spacing and depth can result in similar drainage volume and NO_3^- losses; however, wider spacing could reduce yields due to prolonged water–stress condition (Table 1). In this study, simulation results revealed that for a tile drain spacing of 40 m, reducing the drain depth from 1.5 to 0.9 m reduced NO_3^-–N losses by 31%, whereas for tile drain depth of 1.5 m, increasing tile drain spacing from 27 to 40 m reduced NO_3^-–N losses by 50% (Table 2). In a 6-yr field study in west central Indiana, Hofman et al. (2004) compared 10-,

Table 1. Soil textural class controls hydraulic conductivity and average tile depth and spacing (Schwab et al. 1993).

Soil textural class	Hydraulic class	Hydraulic conductivity	Spacing	Depth
		cm/h	──────── m ────────	
Clay	very slow	0.004	9–15	0.9–1.1
Clay loam	slow	0.004–0.020	12–21	0.9–1.1
Average loam	moderately slow	0.020–0.083	18–30	1.1–1.2
Fine sandy loam	moderate	0.083–0.25	30–37	1.2–1.4
Sandy loam	moderately rapid	0.25–0.54	30–60	1.2–1.5

20-, and 30-m tile-spacing effects on corn production, drainage volume, and NO_3^- loss under silty clay loam soil. Six-year-average highest corn yield and lowest daily NO_3^- concentrations were observed under 20-m spacing, indicating that 30 m might provide insufficient drainage, whereas 10 m might represent an over-drained condition.

Installation of water control structures in tile drainage system provide an opportunity to control drainage volume by varying the drainage depth (Fig. 5a). Control structure conserves soil moisture by retaining water in the soil profile and reduces seasonal and annual drainage volume (Strock et al., 2011). The control structure is left open before planting, allowing free drainage. During crop growth, drainage depth level is raised back to reduce drainage volume and NO_3^- loss (Fig. 5b). However, installing control structure leads to shallow water table depth that may reduce infiltration and increase surface runoff and NO_3^- loss. Drury et al. (2014) found that control structure was effective in reducing tile drainage volume by 9 to 28%, FWMC-NO_3^- by 15 to 33%, and cumulative NO_3^- loss by 38 to 39% relative to unrestricted tile drainage (without control structure). Moreover, Tan et al. (1998) reported an additional advantage of control

structure under no-tillage. A controlled drainage system reduced tile nitrate loss in no-tillage from 38.6 to 28.7 kg N ha[-1] and in a conventional tillage site from 26.9 to 23.1 kg N ha[-1].

Management Practices to Reduce Nutrient Loss under Tile-Drained Conditions

Nutrient losses can be reduced by adoptions of one or more fertilizer management practices like source, rate, placement, application timing, and inhibitors that maximize nutrient use efficiency of crops (Strock et al., 2007). For inorganic-organic N management, monitoring soil inorganic N availability to better synchronize the amount of available N with crop needs and managing soil N availability before, during, and after peak crop N demands are the keys to reducing NO_3^- leaching potential (Dinnes et al., 2002). Fall-applied fertilizer-N are more prone to NO_3^- leaching losses than spring application. There is lack of consensus among researchers regarding the effectiveness of sidedress and split applications to reduce nitrate leaching compared with preplant applications (Randall and Goss, 2008).

Table 2. Predicted (50-yr) changes influence on percent crop yield, drainage (mm), and nitrate losses (kg/ha) under poorly drained clay in southern Minnesota (Nangia et al. 2010).

Tile spacing	Tile depth	Relative yield	Drainage	NO_3–N losses
m	m	%	mm	kg/ha
27	0.9	50	166	21.3
27	1.2	83	202	44.5
27	1.5	100	220	44.9
40	0.9	33	157	15.4
40	1.2	78	177	20.0
40	1.5	93	179	22.3
100	0.9	0	152	11.9
100	1.2	19	152	16.1
100	1.5	46	154	17.9

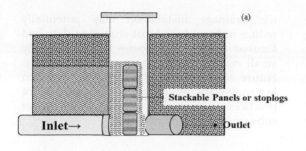

(a)

Stackable Panels or stoplogs

Inlet→

Outlet

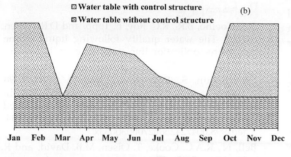

☑ Water table with control structure

☑ Water table without control structure

(b)

Jan Feb Mar Apr May Jun Jul Aug Sep Oct Nov Dec

Fig. 5. (a)Schematic of the water control structure for the tile drainage system and (b) shift in water table with and without control structure.

For reducing P in subsurface drainage, placement of P fertilizer plays a major role. Incorporation of P fertilizer rather than broadcast generally decreases P loss, but incorporation of manure showed slightly greater dissolved P loss than surface application of manure (Ball-Coelho et al., 2007). Differences in P losses in between broadcast and incorporation were greatest in the first precipitation events, but the magnitude of difference was gradually reduced after a couple of rainfall events (King et al., 2014). In regards to P application timing, the chance of loss is more if the time gap in between application time and planting is too long and/or application and the onset of precipitation is short.

Nitrification inhibitors, particularly nitrapyrin, were evaluated to control NO_3^- leaching loss in subsurface drainage, but results were variable depending on soil type and weather pattern (Dinnes et al., 2002). In the upper Midwest, nitrification inhibitors are added to slow the conversion of fall-applied anhydrous NH_3 to the more leachable NO_3^- form (Randall and Goss, 2008). Randall et al. (2003) found that apparent N recovery was 31 and 37% for fall-applied fertilizer-N without and with nitrapyrin, respectively, and 44% for the split (40% preplant in spring and 60% sidedress at V8 stage) N treatment. Randall and Vetsch (2005) concluded that precipitation during May and June had a tremendous control on N recovery. Fertilizer-N management practices like (i) application of nitrapyrin with fall-applied N or spring preplant or (ii) split application would

increase corn yield and profitability under subsurface-drained Mollisols.

Cover crops planted in the fall that uptake residual nitrate and hold it in organic form have been established as effective management to reduce NO_3^- leaching. Cover crop is killed before planting in spring, and trapped N is released to the next crop as cover crop residues mineralized. Cover crops reduced the mass and concentration of leached N by 20 to 80% compared with no cover crop (Meisinger et al., 1991). Drury et al. (2014) found that winter wheat cover crop reduced the 5-yr total tile drainage volume by 7 to 37%, FWMC-NO_3^- concentration in tile drainage water by 21 to 38%, and cumulative NO_3^- loss by 14 to 16% compared with no cover crop. They also concluded that cover crop practice supplemented with tile control structure enhanced the potential of reducing NO_3^- leaching loss by cover crop in cool and humid agricultural soils. The potential of cover crop to sequester inorganic N depends on the species; grasses and brassicas were two to three times more effective than legumes (Dinnes et al., 2002). Cover cropping with rye (*Secale cereale*) has been used successfully in the northern corn and soybean belt. However, cover crop establishment and control of cover crop after spring regrowth were the main obstacles in adoption of cover crop (Strock et al., 2004). Research trials are needed to better understand N releases during cover crop residue decomposition and its interactions with soil type and tillage practices under subsurface drainage conditions.

Conservation tillage practices maintain and enhance preferential flow pathway and increase the amount of drain flow, whereas because of the higher disturbances and mineralization rate, the conventional tillage during early spring can result in early N mineralization peak and increase leachate NO_3^--N concentrations before crops assimilate N (Dinnes et al., 2002). Tan et al. (1998) found that total NO_3^- losses to be 33.7 kg N ha^{-1} for a no-till site and 25.0 kg N ha^{-1} for a conventional tillage site, but the no-till site had 45% higher tile drainage volume than conventional in Brookston clay loam soils. Randall and Irgavarapu (1995) assessed the tillage effect on NO_3^- losses to subsurface during an 11-yr (1982–1992) study under a poorly drained Webster clay loam soil at Waseca, MN. They found that mean annual subsurface drain flow was 35 mm higher for no-tillage than conventional, but NO_3^- losses were about 5% higher with conventional tillage treatment. Combinations of lower soil NO_3^--N levels and higher preferential flow with NT than matrix flow result in lower soil NO_3^--N losses in this system.

One fairly new management practice is the use of subsurface denitrifying bioreactors to reduce the amount of NO_3^- in agricultural drainage. Bioreactors are trenches filled with carbon material (woodchips); tile drain water passes through this bioreactor before it enters local water bodies. Under anoxic conditions, denitrifying bacteria in the bioreactor bed metabolize NO_3^- to nitrogen gas (Bell et al., 2015). Bioreactors reduced NO_3^- loads by 55 to 63% of the corn–soybean rotation (Jaynes et al., 2008; Bell et al., 2015). Bioreactors' NO_3^- removal efficiency depend on retention time controlled by the reactor flow rates combined with the design factors of media porosity and bioreactor flow volume. Very low retention time is not sufficient to reduce the dissolved oxygen to create anoxic environment, a prerequisite for denitrification process, but too long retention times can cause other undesirable processes, like sulfate reduction and mercury methylation (Christianson et al., 2012). It is important to determine the longevity of these bioreactors to justify the cost of bioreactor ($7000 to $9000) and its installation (Christianson et al., 2012).

Designing a nutrient management plan for tile-drained land is critical to reduce nutrient loss from the field and its subsequent negative effect on the environment. It is clear from the above discussion that combinations of multiple management practices suited to weather pattern, rotation, soil type, and growers' economic profitability need to be evaluated to reduce nutrient drainage through subsurface drainage. Including control structure during tile drainage installation may potentially reduce nutrient loss. Most studies to date had focused mostly on N losses and a relatively small number on P movement and drainage. Future research should investigate more on innovative tile design and crop management practices to control N and P losses under subsurface-drained land.

References

Ball-Coelho, B.R., R.C. Roy, E. Topp, and D.R. Lapen. 2007. Tile water quality following liquid swine manure application into standing corn. J. Environ. Qual. 36:580–587 . doi:10.2134/jeq2006.0306

Beauchemin, S., R.R. Simrad, and D. Cluis. 1998. Forms and concentration of phosphorus in drainage water twenty-seven tile-drained soils. J. Environ. Qual. 27:721–728. doi:10.2134/jeq1998.00472425002700030033x

Bell, N., R.A.C. Cooke, T. Olsen, M.B. David, and R. Hudson. 2015. Characterizing the performance of denitrifying bioreactors during simulated subsurface drainage events. J. Environ. Qual. 44:1647–1656. doi:10.2134/jeq2014.04.0162

Burkart, M.R., and D.E. James. 1999. Agricultural-nitrogen contributions to hypoxia in the Gulf of Mexico. J. Environ. Qual. 28:850–859.

Busman, L., and G. Sands. 2002. Agricultural drainage publication series: Issues and answers. Minnesota Ext. Bull. BU-07740-S. Univ. of Minnesota, St. Paul.

Christianson, L.E., A. Bhandari, and M.J. Helmers. 2012. A practice-oriented review of woodchip bioreactors for subsurface agricultural drainage. Appl. Eng. Agric. 28(6):861–874. doi:10.13031/2013.42479

Cooley, E.T., M.D. Ruark, J.C. Panuska. 2013. Tile drainage in Wisconsin: Managing tile-drained landscapes to prevent nutrient loss (GWQ064). http://learningstore.uwex.edu/Assets/pdfs/GWQ064.pdf (accessed 4 Feb. 2016).

Dinnes, D.L., D.L. Karlen, D.B. Jaynes, T.C. Kasper, J.L. Hatfield, T.S. Colvin, and C.A. Cambardella. 2002. Nitrogen management strategies to reduce nitrate leaching in tile-drained Midwestern soils. Agron. J. 94:153–171. doi:10.2134/agronj2002.0153

Drury, C.F., C.S. Tan, T.W. Welacky, W.D. Reynolds, T.Q. Zhang, T.O. Oloya, N.B. McLaughlin, and J.D. Gaynor. 2014. Reducing nitrate loss in tile drainage water with cover crops and water-table management systems. J. Environ. Qual. 43:587–598. doi:10.2134/jeq2012.0495

Fausey, N.R., L.C. Brown, H.W. Belcher, and R.S. Kanwar. 1995. Drainage and water quality in Great Lakes and Cornbelt states. J. Irrig. Drain. Eng. 121(4):283–288. doi:10.1061/(ASCE)0733-9437(1995)121:4(283)

Gburek, W.J., E. Barberis, P.M. Haygarth, B. Kronvang, and C. Stamm. 2005. Phosphorus mobility in the landscape. In: J.T. Sims and A.N. Sharpley, editors, Phosphorus: Agriculture and the environment. Agron. Monogr. 46. ASA, CSSA, and SSSA, Madison, WI. p. 941–979.

Goolsby, D.A., W.A. Battaglin, G.B. Lawrence, R.S. Artz, B.T. Aulenbach, R.P. Hooper, D.R. Keeney, and G.J. Stensland. 1999. Flux and sources of nutrient in the Mississippi-Atchafalaya River Basin. (Report of Task Group 3 to the White House Committee on Environment and Natural Resources, Hypoxia Work Group). Fed. Regist. 64:23834–23835.

Hofman, B.S., S.M. Brouder, and R.F. Turco. 2004. Tile spacing impacts on *Zea mays* L. yield and drainage water nitrate load. Ecol. Eng. 23:251–267. doi:10.1016/j.ecoleng.2004.09.008

Jaynes, D.B., T.C. Kasper, T.B. Moorman, and T.B. Parkin. 2008. In situ bioreactors and deep drain-pipe installation to reduce nitrate losses in artificially drained fields. J. Environ. Qual. 37:429–436. doi:10.2134/jeq2007.0279

King, K.W., M.R. Williams, M.L. Macrae, N.R. Fausey, J. Frankenberger, D.R. Smith, P.J. Kleinman, and L.C. Brown. 2014. Phosphorus transport in agricultural subsurface drainage: A review. J. Environ. Qual. 44:467–485. doi:10.2134/jeq2014.04.0163

Meisinger, J.J., W.L. Hargrove, R.L. Mikkelsen, J.R. Williams, and V.W. Benson. 1991. Effect of cover crops on groundwater quality. In: W.L. Hargrove, editor, Cover crops for clean water. Soil and Water Conserv. Soc., Ankeny, IA. p. 57–68.

Minnesota Pollution Control Agency. 2013. Nitrogen in Minnesota surface waters. https://www.pca.state.mn.us/news/report-nitrogen-surface-water (accessed 5 Feb. 2016).

Nangia, V., P.H. Gowda, D.J. Mulla, and G.R. Sands. 2010. Modeling impacts of tile drain spacing and depth on nitrate-nitrogen losses. Vadose Zone J. 9:61–72. doi:10.2136/vzj2008.0158

Nayak, A.K., R.S. Kanwar, P.N. Rekha, C.K. Hoang, and C.H. Pederson. 2009. Phosphorus leaching to subsurface drain water and soil P buildup in a long-term swine manure applied corn-soybean rotation system. Int. Agric. Eng. J. 18:25–33.

Pavelis, G.A. 1987. Farm drainage in the United States: History, status and prospects. In: USDA Misc. Publ. No. 1455. Econ. Res. Serv., USDA, Washington, DC.

Randall, G.W., and M.J. Goss. 2008. Nitrate losses to surface water through subsurface, tile drainage. In: J.L. Hatfield and R.F. Follett, editors, Nitrogen in the environment: Sources, problems, and management. Elsevier, Burlington, MA. p. 145–175.

Randall, G.W., and T.K. Irgavarapu. 1995. Impact of long-term tillage systems for continuous corn on nitrate leaching to tile drainage. J. Environ. Qual. 24:360–366. doi:10.2134/jeq1995.00472425002400020020x

Randall, G.W., and J.A. Vetsch. 2005. Corn production on a subsurface-drained mollisol as affected by fall versus spring application of nitrogen and nitrapyrin. Agron. J. 97:472–478. doi:10.2134/agronj2005.0472

Randall, G.W., J.A. Vetsch, and J.R. Huffman. 2003. Corn production on a subsurface-drained Mollisol as affected by time of nitrogen application and nitrapyrin. Agron. J. 95:1213–1219. doi:10.2134/agronj2003.1213

Robertson, D.M., G.E. Schwarz, D.A. Saad, and R.B. Alexander. 2014. Incorporating uncertainty into the ranking of SPARROW model nutrient yields from the Mississippi/Atchafalaya River Basin watersheds. National Water-Quality Assessment (NAWQA) Program. USGS. http://water.usgs.gov/nawqa/sparrow/nutrient_yields/ (accessed 5 Feb. 2016).

Sands, G.R. 2001. Agricultural drainage publication series: Soil water concepts. Minnesota Ext. Bull. BU-07644-S. Univ. of Minnesota, St. Paul, MN.

Schwab, G.O., D.D. Fangmeier, and W.J. Elliot. 1996. Soil and water management systems. 4th ed. John Wiley & Sons, New York.

Schwab, G.O., D.D. Fangmeier, W.J. Elliot, and R.K. Frevert. 1993. Soil and water conservation engineering. 4th ed. John Wiley & Sons, New York.

Sims, J.T., R.R. Simard, and B.C. Joern. 1998. Phosphorus loss in agricultural drainage: Historical perspective and current research. J. Environ. Qual. 27:277–293. doi:10.2134/jeq1998.00472425002700020006x

Strock, J.S., C.J. Dell, and J.P. Schmidt. 2007. Managing natural processes in drainage ditches for nonpoint source nitrogen control. J. Soil Water Conserv. 62(4):188–196.

Strock, J.S., P.M. Porter, and M.P. Russelle. 2004. Cover cropping to reduce nitrate loss through subsurface drainage in the northern U.S. Corn Belt. J. Environ. Qual. 33:1010–1016. doi:10.2134/jeq2004.1010

Strock, J.S., G.R. Sands, and M.J. Helmers. 2011. Subsurface drainage design and management to meet agronomic and environmental goals. In: J.L. Hatfield and T.J. Sauer, editors, Soil management: Building a stable base for agriculture. ASA and SSSA, Madison, WI.

Tan, C.S., C.F. Drury, M. Soultani, J.J. van Wesenbeeck, H.Y.F. Ng, J.D. Gaynor, and T.M. Welacky. 1998. Effect of controlled drainage and tillage on soil structure and tile drainage nitrate loss at the field scale. Water Sci. Technol. 38(4-5):103–110. doi:10.1016/S0273-1223(98)00503-4

Tonitto, C., M.B. David, C. Li, and L.E. Drinkwater. 2007. Application of the DNDC model to tile-drained Illinois agroecosystems: Model comparison of conventional and diversified rotations. Nutr. Cycl. Agroecosyst. 78:65–81. doi:10.1007/s10705-006-9074-2

Troeh, F.R., J. Aurthur Hobbs, and R.L. Donahue. 2004. Soil and water conservation for productivity and environmental protection. 4th ed. Prentice Hall, Upper Saddle River, NJ.

Wright, J., and G. Sands. 2001. Agricultural drainage publication series: Planning an agricultural subsurface drainage system. Minnesota Ext. Bull. BU-07685-S. Univ. of Minnesota, St. Paul.

Zhang, T.Q., C.S. Tan, Z.M. Zheng, and C.F. Drury. 2014. Tile drainage phosphorus loss with long-term consistent cropping systems and fertilization. J. Environ. Qual. 44:503–511. doi:10.2134/jeq2014.04.0188

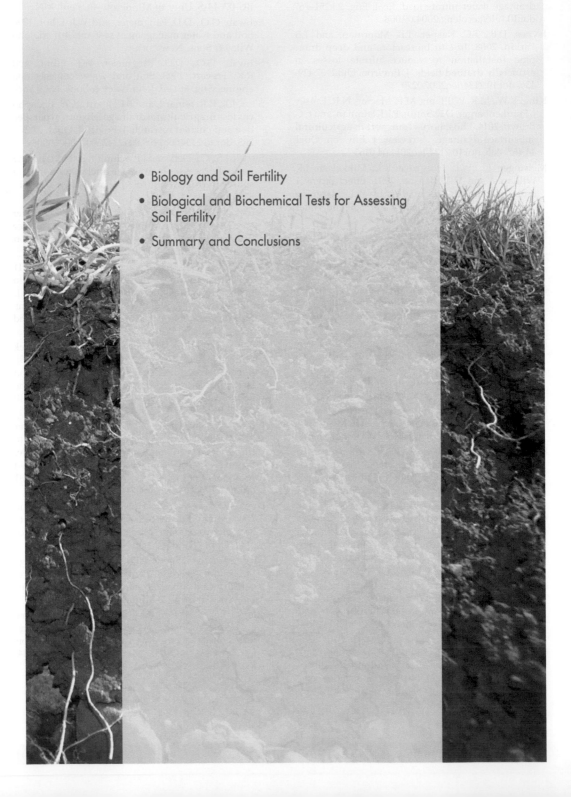

- Biology and Soil Fertility

- Biological and Biochemical Tests for Assessing Soil Fertility

- Summary and Conclusions

Biological and Biochemical Tests for Assessing Soil Fertility

Warren A. Dick* and Steven W. Culman

Assessing soil fertility has, as one of its primary goals, the ability to predict nutrient availability and uptake by a growing crop. It must be understood that there is not always a direct relationship between the soil's fertility level and the response of the crop to that fertility. This is because soil fertility must take into account the many factors of soil chemistry, physics, and biology and how these factors interact with each other and with the weather that occurs during crop growth. Crops are biological systems and soil fertility tests, based upon the soil's biology along with its biological components, provide a link between the soil's potential to supply nutrients and the crops demand for those nutrients. If properly matched, the use efficiency of supplemental nutrients can be increased. This review provides a current assessment of some of the biological and biochemical tests that are being developed and tested to assess soil fertility. Some of these tests are quite well developed and others are more speculative. In all cases, however, the focus is on letting the biology, along with its biochemical components, report to us the amount of nutrients released and the rate of that release to the growing crop.

Biology and Soil Fertility

Broadly speaking, there have been three eras in the production of domesticated crops. The first era arose at the very beginning of agriculture, as mechanical processes were the primary means of soil preparation, weed control, and harvesting. Crude tillage and harvesting tools, some as simple as a sharpened stick to make a hole in the ground to insert seed, were used. As agriculture developed, animal power as well as other more advanced tillage and irrigation tools were incorporated into the production of crops.

During the latter part of the 19th century, the chemical era was initiated. This era's focus was to apply engineered chemical inputs to agricultural systems. It was especially accelerated after World War II when synthetic fertilizers and pesticides became more widely available. The chemical era dominated crop production

Abbreviations: AcdP, acid phosphatase; AlkP, alkaline phosphatase; POXC, permanganate oxidizable C; SOM, soil organic matter.

Warren A. Dick and Steven W. Culman (culman.2@osu.edu), The Ohio State University, The Ohio Agricultural Research and Development Center, 1680 Madison, Ave., Wooster, OH 44691. *Corresponding author (dick.5@osu.edu).

doi:10.2134/soilfertility.2014.0007

© ASA, CSSA, SSSA, 5585 Guilford Rd., Madison, WI 53711-5801, USA.
Soil Fertility Management in Agroecosystems
Amitava Chatterjee and David Clay, Editors

for ~100 yr or more. We have now entered a third era that may be referred to as the biological era. This era can be defined as applying an ever-increasing understanding of the biology of plants, animals, and microorganisms to develop technologies that have further increased crop productivity. Biological approaches may be defined as technologies used to either replace or complement mechanical and chemical technologies. The use of improved seeds and biological seed treatments are examples of this current emphasis on biology. There is certainly overlap in the application of various mechanical, chemical, and biological technologies for enhancing crop production.

Soil fertility can be defined as the capacity of soil to make available sufficient nutrients and moisture to produce crops. The goal of the crop producer is to simultaneously manage a soil that leads to the highest possible crop yields while maintaining long-term sustainability. This requires tools to effectively measure nutrient status in soil. There is a long history within the agronomic sciences to develop soil tests that relate laboratory or simple field tests to actual crop performance. The vast majority of these tests are based on chemical extractions of soil. Fertility tests that are biologically (and biochemically) based may be considered a closer approximation of what the crop is actually experiencing in the field. This is because soil biology and biochemistry are integrated measures of the soil's physical, chemical, and biological properties.

Biological and Biochemical Tests for Assessing Soil Fertility

The biological activity triangle (Fig. 1) illustrates how location, concentration, and target must all come together at the same time for expression of a specific function. For example, for a biological fertility test to be meaningful, it must come from a soil that has been properly sampled to represent a particular location, it must accurately assesses nutrient concentration at that location, and it must be related to the target crop.

A biological soil fertility test must not only provide information that integrates the variables described in the triangle, but there must also be consideration of timing. It does not benefit a crop when there are nutrients available at the right place and right concentration but at a time when there is no crop demand.

In addition, a biological-based soil fertility test must also meet the same criteria as a chemical fertility test in terms of analytical and functional utility. Ideally, the test should be rapid, inexpensive, and properly predict some sort of crop response as a result of nutrient excess or limitation. Chemical tests of soil fertility have been available and used for decades. There is an extensive data set that can relate the test results to a recommendation. Even so, there is much room for improvement, as often the relationship between laboratory tests and field results are poor. This is not surprising because the production of a crop is complex and there are many biotic and abiotic parameters that affect this relationship.

Almost all soil tests are merely a window into the soil at the time the sample is taken. In the future, it may be possible to develop ongoing, real-time assessments of the biological soil fertility based on sensors. These biological tests have the advantage, as previously stated, of integrating the chemical, physical, and biochemical activities in the soil.

After a biological soil fertility test is identified, the next step is to calibrate these tests with crop response field trials so that meaningful recommendations can be developed. This is where research is badly needed, as very little work has been done to connect what is measured in the laboratory with what actually happens in the field over an entire cropping season. Also, there are several challenges to this approach with biological tests that often do not apply to chemical tests. For example, when developing fertilizer recommendations, field trials are routinely established with different plots receiving different rates of fertilizer. It is simple enough to apply multiple rates of mineral fertilizer to plots, but how is this accomplished with soil biology? Attempts to manipulate levels of soil microbial biomass, or metabolic potential, are nearly always wrought with confounding effects in doing so. One approach could be to establish field trials with treatments known to

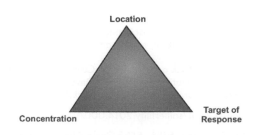

Fig. 1. Biological activity triangle.

influence biological parameters and calibrate identified soil biological tests against this gradient of treatments.

Obtaining a soil sample for a biological soil fertility test may also need to be reexamined. The time of the year and the soil layers sampled may differ from the traditional methods. Soil biology is generally considered to be more variable than soil chemistry—temporally and spatially—and this is especially so when plant residues are added to soil at harvest and biological activity is stimulated for a time.

The goal of this paper is to review and propose new and novel biological tests that can be applied to a soil to guide the management of the soil's fertility. These tests may be only poorly defined at present and will likely require additional work so they can be practically used to guide management. In spite of these difficulties, the development of biological soil fertility tests continues to be of interest, and the next sections provide examples of tests to predict specific soil fertility parameters.

pH

One of the most important properties of soil that affects its fertility is pH. It is not difficult to measure soil pH using a glass electrode or some other device. However, the interpretation of the laboratory measurement requires that experiments be conducted to relate the laboratory results with crop responses, which then lead to lime application recommendations. Dick et al. (2000) proposed a new enzyme approach to assess whether a soil has proper pH balance to provide for good crop growth. An enzyme approach to determine whether a soil is out of balance in terms of pH may be especially appropriate for cropping systems that rely heavily on biological processes to maintain productivity. This approach is based on observations that acid phosphatase predominates in acid soils and alkaline phosphatase in neutral to alkaline soils (Fig. 2). The concept is that there exists an optimum ratio of alkaline phosphatase (AlkP) to acid phosphatase (AcdP) for soils to properly function.

In the initial work that introduced this concept, five acid soils, which varied widely in selected properties, were treated with $CaCO_3$ at rates of 0, 0.2, 0.5, 1.0 and 2.0 times the soil's lime requirement needs (Dick et al., 2000). To remove soil variations in absolute enzyme activity values, an AlkP/AcdP activity ratio was used to test soil response. The ratios of AlkP/

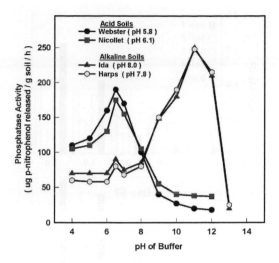

Fig. 2. Acid and alkaline phosphatase activity as affected by soil pH (adopted from Eivazi and Tabatabai, 1977).

AcdP responded immediately to the changes in pH caused by $CaCO_3$ additions (Fig. 3) with some soils yielding higher ratios than others. Incubation of the $CaCO_3$–treated soils for 67 d increased the AlkP/AcdP ratios for two of the soils to a value of 1.5 or higher. However, if we use the 0-d data and an AlkP/AcdP ratio of 1.0 as a first approximation of a proper response to the $CaCO_3$ addition; the results (Fig. 3) suggest that for three of the soils, the 1.0 lime requirement rate was sufficient to adjust soil pH. Two of the soils (Soils 3 and 4) still exhibited a relatively high rate of AcdP activity compared with AlkP activity, suggesting that, for these soils, an additional amount of lime may be needed to properly adjust soil pH and the overall biochemistry of the system.

Additional work is undoubtedly needed to assess soil pH adjustment needs using an enzymatic approach. However, the enzymatic approach may be more accurate than chemical approaches in assessing effective soil pH especially when soils are amended with organic materials. For example, when organic amendments were added to acid soils, the AlkP/AcdP ratio indicated that the need for $CaCO_3$ was reduced (Dick et al., 2000).

Measuring the AlkP/AcdP activity ratio to determine proper pH status of soil seems applicable to a wide variety of soils with different properties and different absolute values of individual phosphatase activities. This is because it is the ratio, not absolute values, that predicts whether a soil is out of pH balance. It is also

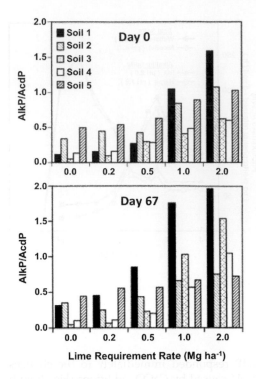

Fig. 3. Alkaline phosphatase/acid phosphatase (AlkP/AcdP) activity ratios at Day 0 and Day 67 after adjusting pH with $CaCO_3$ according to the lime requirement of each soil (Dick et al., 2000).

possible to use the AlkP/AcdP ratio approach on both air-dried or field moist soil samples.

Fertility Assessment Based on Measuring Carbon Fractions in Soil

Soil organic matter (SOM) provides multiple benefits for soils, and the role and management of SOM for creating sustainable farming systems has long been recognized. However, SOM is a broadly defined term, and organic matter that is highly stable and resistant to decay (recalcitrant) is rather inert in directly supporting biological activity.

Lack of attention to maintaining a properly cycling soil organic C fraction can be overcome to some extent by application of irrigation water and high amounts of fertilizer nutrients. For example, yield levels of nonlegume crops in Germany and Switzerland were positively correlated with SOM levels, but the correlation was significant only under conditions of organic farming and not with conventional farming treatments (Brock et al., 2011). However,

it seems reasonable that for most agronomic systems, attention must be paid to proper maintenance of SOM for long-term fertility and sustainability.

The labile C fraction is much more active than the total SOM pool and has a turnover time in soil of months to only a few years. In tropical soils with relatively low levels of inputs, increasing amounts of labile C were found to be associated with higher grain yields (Fig. 4).

Corn (*Zea mays* L.) grain and vegetative biomass were also found to be strongly related to labile soil C and N fractions (Culman et al., 2013). Carbon mineralization (vs. readily oxidizable C, N mineralization, and soil inorganic N) was the best predictor of agronomic performance both individually ($r = 0.61-0.78$, depending on corn stage) and when modeled with multiple indicators.

Several methods to assess the fertility-relevant fractions of SOM have been proposed. One method employs varying concentrations of $KMnO_4$ to separate SOM with different degrees of lability (Loginow et al., 1987). A simplification of the original technique (Lefroy et al., 1993) was used to develop a C management index to describe changes in labile and total C under particular management systems relative to that in an adjoining reference sites (Blair et al., 1995). This method was further developed by Weil et al. (2003) and has been investigated in more detail by Culman et al. (2012), who showed that permanganate oxidizable C (POXC) was sensitive to changes in soil management and that it was strongly related to small and dense particulate organic matter fractions. These particulate fractions are often associated with large amounts of microbial biomass and lower C/N ratios relative to large and light particulate fractions. The interest in this method is due to its analytical simplicity and its ability to positively track management practices that lead to increases in soil C concentrations. The POXC method can be performed under field conditions without any instrumentation at all by simply referring to a color chart. This makes the test appealing for use in both developed and developing countries.

Another method that has gained a large amount of traction in recent years is to estimate readily available mineralizable C via short-term soil respiration measurements. This method measures the amount of carbon dioxide (CO_2) released from rewetted soils in a 24-h (or 3-d) period. These CO_2 tests, for example, the Haney test or the Solvita test, can be used in the field or in the laboratory. Numerous kits, with

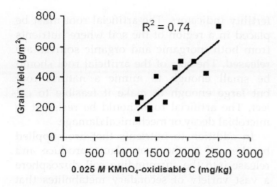

Fig. 4. The relationship between the concentration of labile (oxidizable) C in soil and grain yield in Honduras (Stine and Weil, 2002).

variations of how the CO_2 is measured, have been developed.

These simple CO_2 measurement methods relate well with other biologically active SOM fractions (Franzluebbers et al., 2000), and short-term mineralizable-C has been very strongly correlated with long-term ($\geq$24 d) mineralizable-C rates across many soil types (Franzluebbers et al., 2000; Haney et al., 2001, 2008). Numerous studies have also shown that C and N mineralization rates are highly related and that mineralizable C is a sensitive indicator to recent changes in management (Franzluebbers et al., 2000; Haney et al., 2001; Franzluebbers and Stuedemann, 2008; Schomberg et al., 2009; Vahdat et al., 2010; Culman et al., 2013).

The Haney–Solvita test has been used as an indicator of soil quality and soil health. There is often a positive relationship between the amount of CO_2 measured by the Solvita test and improvement in soil quality resulting from cover crops, organic manure additions,

residue additions, reduced tillage, etc. The Haney–Solvita test also is considered an indicator of soil nutrient availability because as soil organic matter is mineralized and releases CO_2, it also releases plant available nutrients such as N, P, and S. However, CO_2 emissions from soil, assumed to be primarily a result of microbial respiration, are highly variable and the factors controlling respiration are not always clearly understood. Work conducted in Minnesota showed only a weak relationship between a short-term CO_2 burst and the economically optimum N rate and corn grain yield (Fig. 5; Tu et al., 2014, 2015). This is not surprising, as there are many other variables that also affect crop growth and yield in addition to N supplied by the soil. There is clearly a need to further develop the Haney–Solvita test for measuring soil biological properties and functions and relating these to crop yields for various soils and climates (Sullivan, 2015). In the study by Culman et al. (2013), simple organic matter tests were compared in a long-term systems trial in Michigan. They linked soil properties and corn growth over the course of the growing season. Early-season (preplant, fifth leaf, and 10th leaf) mineralizable C was found to be the best predictor of corn grain yield and total aboveground biomass, bettering the two recommended methods for early-season corn N status, i.e., the presidedress nitrate test (Magdoff et al., 1984), and the leaf chlorophyll content (Piekkielek and Fox, 1992).

The microbial biomass, another C fraction that makes up ~2 to 3% of the total soil organic C, is a tremendous storehouse of energy and nutrients and controls their flow through soil. The soil microbial biomass can only be sustained as a healthy, functioning fraction of soil if organic matter (i.e., plant residues and animal

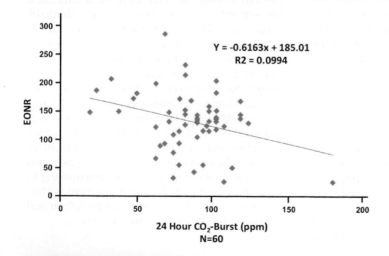

Fig. 5. Correlation between 24-h Solvita CO_2-burst (ppm) and economically optimum N fertilizer rate (EONR) (adopted from Tu et al., 2014).

manures) is introduced into the soil on a regular basis. The continuous cycling of this organic matter creates a microbial diversity and balance that helps control insect and disease pests. It is also important in cycling nutrients back to the crop. Very stable or recalcitrant C only poorly supports a highly functioning microbial biomass and long-term soil fertility.

The size and the turnover rate of the microbial biomass are very important in determining how efficiently nutrients are cycled, and methods are needed that can rapidly assess both microbial biomass size and turnover. Ideally, the microbial biomass should be large enough and turn over fast enough to supply sufficient nutrients to the crop at the proper time needed to support good crop growth and yield.

A rapid microwave method to assess the soil microbial biomass (both C and N biomass) has been developed. The high temperature and vapor pressure created by the microwave energy affects the permeability and stability of the cell membrane, thus causing mechanical rupture of the cell (Islam and Weil, 1998). The difference in soluble C (and N) in soils with or without microwave treatment thus is a reflection of the microbial biomass size. In many studies, there is a strong relationship between the microwave method and the more traditional chloroform fumigation method. de Araujo (2010) indicated the necessity to calibrate the microwave method for different soils with a range of properties, such as clay content, to find an appropriate conversion factor to generate correct values for calculating the soil microbial biomass.

It must be noted that at an even more basic level is the biochemistry expressed by the microbial biomass. This involves expression of genes and enzyme activities in support of a specific soil function, for example, amino acid mineralization to produce ammonia. Work is only beginning to assess the relevant genes in the microbial biomass that relate to soil fertility and long-term soil quality (sustainability).

Artificial Roots and Ion Exchange Membranes

Using artificial roots as indicators of what is happening in the soil is an integrative method that holds promise both as a one-time measure of soil fertility or for continuous sampling throughout a growing season. Martens (1982) proposed several criteria for development of artificial roots that pertain to their use as soil fertility indicators. The artificial root must be placed in a region of the soil where nutrients from both inorganic and organic sources are released. The size of the artificial root should be small enough to mimic a natural root but large enough to make it feasible to collect. The artificial root should be resistant to microbial decay or mechanical damage.

In addition to nutrients that are supplied by the soil to the root, plants produce and release into the surrounding soil rhizosphere a vast variety of secondary metabolites that can also affect plant growth and crop production. For example, allelopathic compounds may be released from the plant to the soil to restrict competition by other plants for nutrients and water. A silicone tube microextraction technique was developed by Mohney et al. (2009) to measure nonpolar root exudates, thus providing a means to test hypotheses about the role of roots in plant–plant interactions. The sampling method is quite simple, the materials are inexpensive, and it is sensitive to daily and spatial variations in chemical content. In one example, a 1-m length of polydimethylsiloxane microtubing was placed in soil without disturbance of plant roots. The two ends of the tubing remained out of the soil so that solution could be washed through the tubing to collect samples for analysis. This technique should also be applicable to the measurement of more polar root exudates, thus providing a way to measure nutrient release throughout the growing season. Indeed, a technique that is similar, but at a macroscale, has been developed for measuring soil P and K (Yang et al., 2012).

Approximately 20 to 25% of the N applied to the field as fertilizer and manure is lost to the environment from heavy soils via denitrification (van der Salm et al., 2007). After heavy rainfall events that can leave soils saturated for several days, the loss of N via denitrification may be even greater (Laboski, 2008). Because N is such an important nutrient, a way to measure real-time estimates of N loss via denitrification would be a valuable tool to assist crop producers. Although much larger in scale than microtubes described by Mohney et al. (2009), regular laboratory silicone tubing can be used for sampling of soil gases (Jacinthe and Dick, 1996). After placement in soil, it takes several hours for the gases outside of the tube (but in the soil atmosphere) to equilibrate with what is sampled from within the tube. The silicone tubes can be left in soil for making repeated measurements.

Root probe simulators using ion exchange resins can also measure plant nutrient supplies in soil. Generally, the ion exchange resins are embedded in a solid support to provide stability and ease of use as a soil probe (Fig. 6). One advantage of such exchange resins is that multiple elements can be simultaneously extracted and measured. In a procedure reported by McLaughlin et al (1993), both cations and anions were extracted by shaking soil with exchange resins in distilled water. The resins are separated from the soil, the ions are desorbed from the resins, and the concentrations of desorbed ions are then determined by conventional means. The results of the resin extractions correlate well with standard chemical extraction procedures (Fig. 7). Alternatively, probes can be left in the soil for hours to days before removing them and then measuring cations and anions that were transferred from the soil solution to the resins. These probes are convenient to use and economical and have been evaluated in a number of nutrient assessment and field fertility studies (Szillery et al., 2006; Qian and Schoenau, 2005; Askegaard et al., 2005; Salisbury and Christensen, 2000; Subler et al., 1995).

Active Fractions of Nutrients

Almost all the N in soil occurs in organic forms. Generally, the N in these organic forms has to be mineralized to inorganic N forms before being taken up by plants. The C/N ratio in soil is ~12:1, but not all of the organic material in soil is easily mineralized. There are some fractions that cycle much more rapidly. Tests have been proposed that look at the amino sugar and the amino acid fractions of the soil to predict N fertility status. The supply of N from these fractions to the crop depends on their pool size and turnover rate. However, even with efficient cycling, sufficient N may not be available at the time the crop most needs it. In this case, a measurement of how much N is potentially available from the organic sources in soil can be used as a baseline for soil fertility recommendations, and additional fertilizer N can be added as supplement for optimum crop production.

A soil test that measures N released from soil after being heated with alkali (2 M NaOH) provides an estimate of amino sugar concentrations in soil (Mulvaney and Khan, 2001; Khan et al., 2001). This test was proposed as a way to identify soils that do not respond to N fertilization. A simple near-infrared reflectance spectroscopy

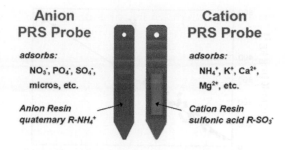

Anion PRS Probe

adsorbs:

NO_3^-, PO_4^-, SO_4^-, micros, etc.

Anion Resin quaternary $R-NH_4^+$

Cation PRS Probe

adsorbs:

NH_4^+, K^+, Ca^{2+}, Mg^{2+}, etc.

Cation Resin sulfonic acid $R-SO_3^-$

Fig. 6. Example of anion and cation probes (Western Ag, 2016).

method has also been developed by Dick et al. (2013) to measure this same fraction.

The concentration of total free amino acids in soil solutions and soil extracts has been reported by Jones et al. (2002). Work by Jones and Kielland (2002), although conducted in Arctic soils and not agricultural soils, has shown that the amino acid pool in the soil solution is extremely transient, turning over ~20 times per day. Thus the transformation of protein or other larger macromolecules in soil, and not amino acids (or presumably other similar soil fractions like amino sugars), may be the factor limiting N availability from organic sources in soil. This would suggest that tests that focus on the metabolism of proteins might be more appropriate to estimate soil N fertility.

Nitrogen fractions assess the potential availability of N that is transformed (i.e., mineralized) from organic forms to inorganic forms that are then taken up by the crop. However, a single N fraction cannot represent all of the organic N pools that contribute to mineral N concentrations in soil. Therefore, studies have been conducted to assess overall N mineralization in soils. The classical method of Stanford and Smith method (Stanford and Smith, 1972) was an advancement in this effort but required long incubation times. Thus, more rapid estimates that can predict soil organic N mineralization are desired.

A method developed by Gianello and Bremner (1986) exposed soil for 4 h to hot (100°C) KCl to measure NH_4^+ release. The greater the amount of NH_4^+ measured, the larger the potentially mineralizable N pool size in soil. Picone et al. (2002) modified and then evaluated another rapid method to measure potentially mineralizable N based on measuring the gas pressure generated when soil is treated with $Ca(ClO)_2$ in a closed vessel. An infrared method to measure amino sugar N (Dick et al., 2013), and

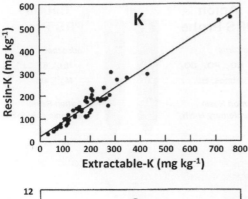

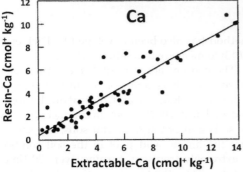

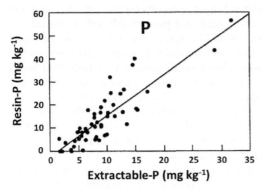

Fig. 7. Relationship between concentrations of K, Ca, and P determined by ion exchange resins and by conventional soil tests for 50 soils (adopted from McLaughlin et al., 1993).

potentially mineralizable N, has also been developed (Moron and Cozzolino, 2004). At first glance, it may seem like the amount of useful data that can be extracted from an infrared scan of a soil sample is insufficient for estimating soil property values. However, the first and second differentials of infrared scans (Fig. 8), along with specialized software, can be used to tease out relationships that are highly predictive of soil properties (Fig. 9).

Infrared methods show great promise because they require minimal soil preparation, the equipment needed is not expensive, it is rapid and the prediction equations can be constantly improved as more data points are entered into the correlations between values of mineralizable and available N measured in the laboratory/field vs. infrared values. The biological fractions impacted by these rapid methods are not known and remain to be identified.

In spite of the poor results to date, the goal remains to develop rapid methods that accurately assess N supply in soil that relate crop production results to a laboratory result. That such a goal has remained elusive is not surprising because intense agricultural production has often reduced the concentration of organic matter that is readily cycled on an annual basis to the point where it is no longer possible for optimum crop production to proceed without supplemental nutrient inputs. However, knowing what fraction of the total N needs of the crop can be supplied by soil organic matter is still important information in improving N-use efficiency. Overall, environmental quality would also improve as more accurate N fertilizer application rates are made so as to avoid excess N being lost as nitrates to water or as nitrous oxide to the atmosphere.

Although soil tests are available for S, they are generally not as precise as for K and P. This is because like N, the majority of S in soil is in organic forms. Thus, S fertility tests based on chemical extractions methods often give poor results. For example, S extracted by $CaCl_2$ after 8 wk of plant growth could not account for plant uptake (Goh and Pamidi, 2003). This strongly suggests S must be originating directly from the mineralization of soil organic S pools. Intensive cropping with continuous use of 100% NPK without S resulted in depletion of total, organic, and inorganic S concentrations by 18.1, 17.8, and 21.7%, respectively (Reddy et al., 2001).

The two major organic forms of S in soil are ester sulfates and C-bonded S (or primarily amino acid S). In S-deficient soils, lower percentages of ester sulfate relative to C-bonded S usually exist than in nondeficient-S soils (Edwards, 1998). Sulfatase enzymes in soil can rapidly release mineral S from ester sulfates, while C-bonded S relies on microbial activity for mineralization. Rapid fertility tests based on (i) the sulfatase reaction (Tabatabai and Bremner, 1970), (ii) a rapid S mineralization reaction test similar to that proposed for N, or (iii) a combination of both approaches would thus seem

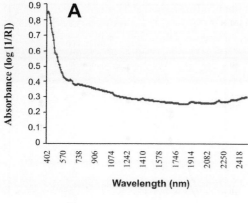

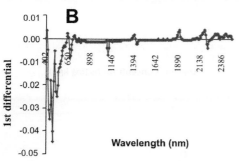

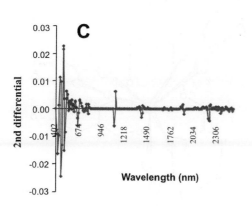

Fig. 8. Spectrum of a soil sample: (A) raw spectral data; (B) first differential of spectral data; and (C) second differential of spectral data (Dick et al. 2013).

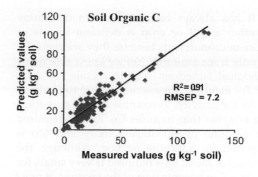

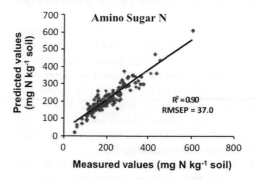

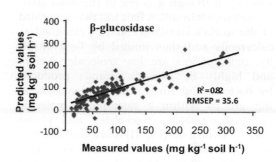

Fig. 9. Plots depicting the relationship of measured and near infrared spectroscopy (NIRS) predicted values (Dick et al. 2013).

promising to determine crop-available S in soil for fertility assessments.

Similarly, microbial activity is known to be important in redistributing P into different forms. However, the organic forms of P are less important for crop nutrition than is the case for S. For example, changes in inorganic and organic P fractions resulting from 65 yr of intensive cropping by a wheat (*Triticum aestivum* L.)–wheat–fallow rotation indicated that only 22% of the P lost came from extractable organic forms (Hedley et al., 1982).

Micronutrient Availability

As we continue to harvest high-yielding crops, we remove ever-increasing amounts of micronutrients. There is concern that we may have reached a point where crop yields are limited by a lack of micronutrients.

It has always been difficult to determine whether a soil or crop is deficient in one or more micronutrients because they are required in only trace amounts. Yet we know that many biological functions require metals for activity. For example, urease requires up to 12 atoms of Ni for activity. Tyrosinase is a humus-forming enzyme that requires Co. Zinc is required for alkaline phosphatase activity and Mo is required for xanthine oxidase. Although the amount of metal ions required is very small, for the metal to become part of the enzyme, it must be available during the synthesis of the enzyme. It may also be required for the activity of the enzyme to be expressed. If these metals are limiting, the activity of the enzymes that require them are also reduced (Yu, 2006).

Preliminary work has shown that biological enzyme assays can be used to determine the bioavailability of Ni and Zn. (Yu, 2006). For example, AlkP activity was found limited by lack of Zn, and when Zn was added to the soil before making an enzyme measurement, AlkP activity was increased (Fig. 10).

Iron is another micronutrient that can become limiting, because of the low solubility of the ferric iron (Fe^{3+}) ion in neutral to alkaline soils, even though it is one of the most abundant elements in soil. A little less than one-third of the world's farmland could be classified as calcareous and thus limited by Fe availability. Siderophores are low molecular weight and highly Fe^{3+} specific ligands produced by bacteria and fungi. The dominant chelating groups within a siderophore molecule are hydroxamates, catecholates, and phenolates. Enterobactin (Fig. 11) is one the strongest known chelators of Fe^{3+}. Its affinity constant (K) is equal to 10^{52} M^{-1}. Although enterobactin has a high affinity for Fe^{3+}, the Fe^{3+} can be rapidly exchanged within the plant and utilized.

Phytosiderophores, produced by plants, have slightly different structures than microbial siderophores, but both are important in supplying Fe to plants in soils with limited Fe availability. Siderophores have also been investigated for their role in biological disease control (Jagadeesh et al., 2001). Several methods have been developed to quantify siderophores in soil (Essén et al., 2006; Bossier and Verstraete, 1986). Little effort has been made to relate siderophore (or phytosiderophore) concentrations in soil to Fe nutrition, but this seems to be an area ripe for development.

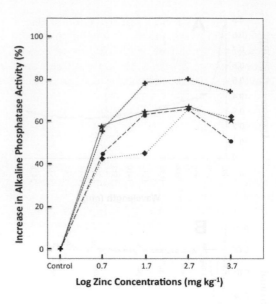

Fig. 10. Stimulation of alkaline phosphatase activity in four soils by addition of Zn.

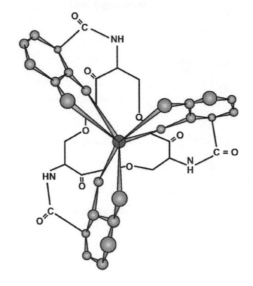

Fig. 11. Structure of enterobactin, a catecholate siderophore with iron at the center of a six coordinate, octahedral complete (Neilands, 1981).

Redox Status of Soil

Aerated soils have redox potentials in the range of +400 to +700 mV. When air (oxygen) is restricted from the soil, most commonly because of water saturation, the redox potential rapidly decreases as chemical species other than oxygen become the terminal electron acceptors. Eh and pH are both factors that strongly influence the availability and mobility of many nutrients in complex chemical and biological environment (Husson, 2013). For example, nitrates are converted to nitrous oxide and dinitrogen gas and lost when soil becomes reduced.

Electrodes can be used to measure the redox status in soil. Whole-soil Eh values have little meaning, however, because redox potentials may vary widely over short distances (Sexstone et al., 1985). Thus electrode measurements cannot adequately define the nutrient status of the soil as affected by the soils redox potential. The presence of various reduced chemical species in soil is often a better indicator of the redox status of a soil than measured Eh values. An alternative method is to use a simple dehydrogenase enzyme assay that reflects and integrates the entire soil volume. Soil dehydrogenase seems to work well for this purpose (Fig. 12). Measurement of dehydrogenase activity can be made quickly, and it is inversely related to the redox status of the soil.

Summary and Conclusions

Biological indicators to aid in the management of soil fertility offer great potential. It also requires that research be conducted to develop this potential. At this time, it is difficult to determine optimal nutrition concentrations in soil for supporting crop production based only on soil biological tests. New tests and further refinement of existing tests are much needed. Once a test is proposed, it must be evaluated across geographical regions and for various crops. For example, effective soil acidity measured using the ratio of AlkP to AcdP will be different for crops. Crops such as alfalfa (*Medicago sativa* L.) generally require near neutral soil pH compared with crops, like blueberry (*Dianella nigra* Colenso), that like more acid soils.

As more emphasis is placed on maintaining or increasing organic matter levels in soil, new biologically based soil fertility tests would seem to have an important role in helping to predict nutrient contributions from the soil to the crop. Can biological assays do a better job than chemical assays in a system where labile C fractions and microbial biomass are actively cycling? Biological soil fertility tests may also provide benefits in systems where a combination of organic and inorganic inputs are applied to soil to increase overall nutrient use efficiency.

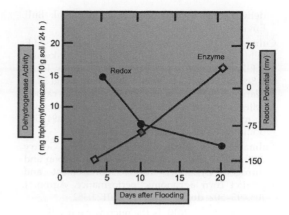

Fig. 12. Relationship between dehydrogenase activity and redox potential in a flooded soil (adopted from Chendrayan and Sethunathan, 1980).

Site-specific management is currently being developed using data from sensors that measure soil properties on a real-time basis. It is proposed that simple sensor tools also be developed that can measure soil biological properties on a real-time basis. These tools would be important contributors to assessing the overall fertility status of the soil at any point in time. Biologically based tools can provide overall integration of the soil's chemistry, physics, and biology as they affect the fertility of a soil. The goal is more efficient nutrient cycling and less negative environmental impacts.

References

Askegaard, M., H.C.B. Hansen, and J.K. Schjoerring. 2005. A cation exchange resin method for measuring long-term potassium release rates from soil. Plant Soil 271:63–74. doi:10.1007/s11104-004-2025-2

Blair, G.J., R.D.B. Lefroy, and L. Lisle. 1995. Soil carbon fractions based on their degree of oxidation, and the development of a Carbon Management Index for agricultural systems. Aust. J. Agric. Res. 46:1459–1466. doi:10.1071/AR9951459

Bossier, P., and W. Verstraete. 1986. Detection of siderophores in soil by a direct bioassay. Soil Biol. Biochem. 18:481–486. doi:10.1016/0038-0717(86)90004-0

Brock, C., A. Fließbach, H.R. Oberholzer, F. Schulz, K. Wiesinger, F. Reinicke, W. Koch, B. Pallutt, B. Dittman, J. Zimmer, K.J. Hülsbergen, and G. Leithold. 2011. Relation between soil organic matter and yield levels of nonlegume crops in organic and conventional farming systems. J. Plant Nutr. Soil Sci. 174:568–575. doi:10.1002/jpln.201000272

Chendrayan K., and N. Sethunathan. 1980. Effects of HCH, carbaryl, benomyl, and atrazine on the dehydrogenase activity of a flooded soil. Bull. Environ. Contam. Toxicol. 24:379–382.

Culman, S.W., S.S. Snapp, M.A. Freeman, M.E. Schipanski, J. Beniston, R. Lal, L.E. Drinkwater, A.J. Franzluebbers, J.D. Glover, A.S. Grandy, J. Lee, J. Six, J.E. Maul, S.B. Mirksy, J.T. Spargo, and M.M. Wander. 2012. Permanganate oxidizable carbon reflects a processed soil fraction that is sensitive to management. Soil Sci. Soc. Am. J. 76:494–504. doi:10.2136/sssaj2011.0286

Culman, S.W., S.S. Snapp, J.M. Green, and L.E. Gentry. 2013. Short- and long-term labile soil carbon and nitrogen dynamics reflect management and predict corn agronomic performance. Agron. J. 105:493–502. doi:10.2134/agronj2012.0382

de Araujo, A.S.F. 2010. Is the microwave irradiation a suitable method for measuring soil microbial biomass? Rev. Environ. Sci. Bio/Technol. 9:317–321. doi:10.1007/s11157-010-9210-y

Dick, W.A., L. Cheng, and P. Wang. 2000. Soil acid and alkaline phosphatase activity as pH adjustment indicators. Soil Biol. Biochem. 32:1915–1919. doi:10.1016/S0038-0717(00)00166-8

Dick, W.A., B. Thavamani, S. Conley, R. Blaisdell, and A. Sengupta. 2013. Prediction of b-glucosidase and b-glucosaminidase activities, soil organic C, and amino sugar N in a diverse population of soils using near infrared reflectance spectroscopy. Soil Biol. Biochem. 56:99–104. doi:10.1016/j.soilbio.2012.04.003

Edwards, P.J. 1998. Sulfur cycling, retention and mobility in soils: A review. General Technical Report NE-250, USDA Forest Service Publications Distribution, Delaware, OH.

Eivazi, F., and M.A. Tabatabai. 1977. Phosphatases in soils. Soil Biol. Biochem. 9:167–172. doi:10.1016/0038-0717(77)90070-0

Essén, S.A., D. Bylund, S.J.M. Holmström, M. Moberg, and U.S. Lundström. 2006. Quantification of hydroxamate siderophores in soil solutions of podzolic soil profiles in Sweden. BioMetals 19:269–282. doi:10.1007/s10534-005-8418-8

Franzluebbers, A.J., R.L. Haney, C.W. Honeycutt, H.H. Schomberg, and F.M. Hons. 2000. Flush of carbon dioxide following rewetting of dried soil relates to active organic pools. Soil Sci. Soc. Am. J. 64:613–623. doi:10.2136/sssaj2000.642613x

Franzluebbers, A.J., and J.A. Stuedemann. 2008. Early response of soil organic fractions to tillage and integrated crop–livestock production. Soil Sci. Soc. Am. J. 72:613–625. doi:10.2136/sssaj2007.0121

Gianello, C., and J.M. Bremner. 1986. Comparison of chemical methods for assessing potentially mineralizable organic nitrogen in soil. Commun. Soil Sci. Plant Anal. 17:215–236. doi:10.1080/00103628609367709

Goh, K.M., and J. Pamidi. 2003. Plant uptake of sulphur as related to changes in the HI-reducible and total sulphur fractions in soil. Plant Soil 250:1–13. doi:10.1023/A:1022823319406

Haney, R.L., W.H. Brinton, and E. Evans. 2008. Estimating soil carbon, nitrogen, and phosphorus minerlization from short-term carbon dioxide respiration. Commun. Soil Sci. Plant Anal. 39:2706–2720. doi:10.1080/00103620802358862

Haney, R., F. Hons, M. Sanderson, and A. Franzluebbers. 2001. A rapid procedure for estimating nitrogen mineralization in manured soil. Biol. Fertil. Soils 33:100–104. doi:10.1007/s003740000294

Hedley, M.J., J.W.B. Stewart, and B.S. Chauhan. 1982. Changes in inorganic and organic soil phosphorus fractions induced by cultivation practices and by laboratory incubations. Soil Sci. Soc. Am. J. 46:970–976. doi:10.2136/sssaj1982.03615995004600050017x

Husson, O. 2013. Redox potential (Eh) and pH as drivers of soil/plant/microorganism systems: A transdisciplinary overview pointing to integrative opportunities for agronomy. Plant Soil 362:389–417. doi:10.1007/s11104-012-1429-7

Islam, K.R., and R.R. Weil. 1998. Microwave irradiation of soil for routine measurement of soil microbial biomass carbon. Biol. Fertil. Soils 27:408–416. doi:10.1007/s003740050451

Jacinthe, P.-A., and W.A. Dick. 1996. Use of silicone tubing to sample nitrous oxide in the soil atmosphere. Soil Biol. Biochem. 28:721–726. doi:10.1016/0038-0717(95)00176-X

Jagadeesh, K.S., J.H. Kulkarni, and P.U. Krishnaraj. 2001. Evaluation of the role of fluorescent siderophore in the biological control of bacterial wilt in tomato using Tn5 mutants of fluorescent Pseudomonas sp. Curr. Sci. 81:882–883.

Jones, D.L., and K. Kielland. 2002. Soil amino acid turnover dominates the nitrogen flux in permafrost-dominated taiga forest soils. Soil Biol. Biochem. 34:209–219. doi:10.1016/S0038-0717(01)00175-4

Jones, D.L., A.G. Owen, and J.F. Farrar. 2002. Simple method to enable the high resolution determination of total free amino acids in soil solutions and soil extracts. Soil Biol. Biochem. 34:1893–1902. doi:10.1016/S0038-0717(02)00203-1

Khan, S.A., R.L. Mulvaney, and R.G. Hoeft. 2001. A simple soil test for detecting sites that are nonresponsive to nitrogen fertilization. Soil Sci. Soc. Am. J. 65:1751–1760. doi:10.2136/sssaj2001.1751

Laboski, C. 2008. Potential for nitrogen loss from heavy rainfalls. http://www.uwex.edu/ces/ag/issues/Nloss.html (accessed 30 Dec. 2014).

Lefroy, R.D.B., G.J. Blair, and W.M. Strong. 1993. Changes in soil organic matter with cropping as measured by organic carbon fractions and ^{13}C natural isotope abundance. Plant Soil 155-156:399–402. doi:10.1007/BF00025067

Loginow, W., W. Wisniewski, S.S. Gonet, and B. Ciescinska. 1987. Fractionation of organic carbon based on susceptibility to oxidation. Pol. J. Soil Sci. 20:47–52.

Magdoff, F.R., D. Ross, and J. Amadon. 1984. A soil test for nitrogen availability to corn. Soil Sci. Soc. Am. J. 48:1301–1304. doi:10.2136/sssaj1984.03615995004800060020x

Martens, R. 1982. Apparatus to study the quantitative relationships between root exudates and microbial populations in the rhizosphere. Soil Biol. Biochem. 14:315–317. doi:10.1016/0038-0717(82)90046-3

McLaughlin, M.J., P.A. Lancaster, P.W.G. Sale, N.C. Uren, and K.I. Peverill. 1993. Use of cation/anion exchange membranes for multi-element testing of acidic soils. Plant Soil 155–156:223–226. doi:10.1007/BF00025024

Mohney, B.K., T. Matz, J. LaMoreaux, D.S. Wilcox, A.L. Gimsing, P. Mayer, and J.D. Weidenhamer. 2009. In situ silicone tube microextraction: A new method for undisturbed sampling of root-exuded thiophenes from marigold (*Tagetes erecta* L.) in soil. J. Chem. Ecol. 35:1279–1287. doi:10.1007/s10886-009-9711-8

Moron, A., and D. Cozzolino. 2004. Determination of potentially mineralizable nitrogen and nitrogen in particulate organic matter fractions in soil by visible and near-infrared reflectance spectroscopy. J. Agric. Sci. 142:335–343. doi:10.1017/S0021859604004290

Mulvaney, R.L., and S.A. Khan. 2001. Diffusion methods to determine different forms of nitrogen in soil hydrolysates. Soil Sci. Soc. Am. J. 65:1284–1292. doi:10.2136/sssaj2001.6541284x

Neilands, J.B. 1981. Microbial iron compounds. Annu. Rev. Biochem. 50:715–731. doi:10.1146/annurev.bi.50.070181.003435

Picone, L.I., M.L. Cabrera, and A.J. Franzluebbers. 2002. A rapid method to estimate potentially mineralizable nitrogen in soil. Soil Sci. Soc. Am. J. 66:1843–1847. doi:10.2136/sssaj2002.1843

Piekkielek, W.P., and R.H. Fox. 1992. Use of a chlorophyll meter to predict sidedress nitrogen. Agron. J. 84:59–65. doi:10.2134/agronj1992.00021962008400010013x

Qian, P., and J.J. Schoenau. 2005. Use of ion-exchange membrane to assess nitrogen-supply power of soils. J. Plant Nutr. 28:2193–2200. doi:10.1080/01904160500324717

. Reddy, K.S., M. Singh, A.K. Tripathi, A. Swarup, and A.K. Dwivedi. 2001. Changes in organic and inorganic sulfur fractions and S mineralisation in a Typic Haplustert after long-term cropping with different fertiliser and organic manure inputs. Aust. J. Soil Res. 39:737–748. doi:10.1071/SR00020

Salisbury, S.E., and N.W. Christensen. 2000. Exchange resins measure rotation effect on nutrient availability. Better Crops 84:14–16.

Schomberg, H.H., S. Wietholter, T.S. Griffin, D.W. Reeves, M.L. Cabrera, D.S. Fisher, D.M. Endale, J.M. Novak, K.S. Balkcom, R.R. Raper, N.R. Kitchen, M.A. Locke, K.N. Potter, R.C. Schwartz, C.C. Truman, and D.D. Tyler. 2009. Assessing indices for predicting potential nitrogen mineralization in soils under different management systems. Soil Sci. Soc. Am. J. 73:1575–1586. doi:10.2136/sssaj2008.0303

Sexstone, A.J., N.P. Revsbech, T.B. Parkin, and J.M. Tiedje. 1985. Direct measurement of oxygen profiles and denitrification rates in soil aggregates. Soil Sci. Soc. Am. J. 49:645–651. doi:10.2136/sssaj1985.03615995004900030024x

Stanford, G., and S.J. Smith. 1972. Nitrogen mineralization potentials of soils. Soil Sci. Soc. Am. Proc. 36:465–472. doi:10.2136/sssaj1972.03615995003600030029x

Stine, M.A., and R.R. Weil. 2002. The relationship between soil quality and crop productivity across three tillage systems in south central Honduras. Am. J. Altern. Agric. 17:1–8.

Subler, S., J.M. Blair, and C.A. Edwards. 1995. Using anion-exchange membranes to measure soil nitrate availability and net nitrification. Soil Biol. Biochem. 27:911–917. doi:10.1016/0038-0717(95)00008-3

Sullivan, D. 2015. Are "Haney Tests" meaningful indicators of soil health and estimators of nitrogen fertilizer credits. WERA-103 Committee Publication. Nutrient Digest 7:1–2.

Szillery, J.E., I.J. Fernandez, S.A. Norton, E. Rustad, and A.L. White. 2006. Using ion-exchange resins to study soil response to experimental watershed acidification. Environ. Monit. Assess. 116:383–398. doi:10.1007/s10661-006-7462-3

Tabatabai, M.A., and J.M. Bremner. 1970. Arylsulfatase activity of soils. Soil Sci. Soc. Am. J. 34:225–229.

Tu, X., D.L. Allan, C. Rosen, J.A. Coulter, and D.E. Kasier. 2015. Evaluation of the Solvita test as an indicator of mineralizable nitrogen in Minnesota soils. Presented at the Annual Meeting of the ASA/CSSA/SSSA, Minneapolis, MN. 15–18 Nov. 2015. https://scisoc.confex.com/scisoc/2015am/webprogram/Paper94351.html (accessed 27 Jan. 2016).

Tu, X., D.L. Allan, C. Rosen and D.E. Kasier. 2014. Solvita and other mineralizable nitrogen tests as indicators of N response in Minnesota soils. Presented at the Annual Meeting of the ASA/CSSA/SSSA, Long Beach, CA. 2–5 Nov. 2014. https://scisoc.confex.com/scisoc/2014am/webprogram/Paper87853.html. (accessed 27 Jan. 2016).

Vahdat, E., F. Nourbakhsh, and M. Basiri. 2010. Estimation of net N mineralization from short-term C evolution in a plant residue-amended soil: Is the accuracy of estimation time-dependent? Soil Use Manage. 26:340–345. doi:10.1111/j.1475-2743.2010.00285.x

van der Salm, C., J. Dolfing, M. Heinen, and G.L. Velthof. 2007. Estimation of nitrogen losses via denitrification from a heavy clay soil under grass. Agric. Ecosyst. Environ. 119:311–319. doi:10.1016/j.agee.2006.07.018

Weil, R.R., K.R. Islam, M.A. Stine, J.B. Gruver, and S.E. Samson-Liebig. 2003. Estimating active carbon for soil quality assessment: A simplified method for laboratory and field use. Am. J. Altern. Agric. 18:3–17. doi:10.1079/AJAA2003003

Western Ag. 2016. Basics. http://www.westernag.ca/innovations/technology/basics (accessed 27 Jan. 2016).

Yang, X.J., Y.L. Lai, J.Y. Mo, and H. Shen. 2012. A device for simulating soil nutrient extraction and plant uptake. Pedosphere 22:755–763. doi:10.1016/S1002-0160(12)60061-7

Yu, S. 2006. Assessment of bioavailable nickel in soil using nickel-depleted urease. MS thesis. The Ohio State University, Columbus, OH.

Stone, M.A., and K.R. Weil. 2002. The relationship between soil quality and crop productivity across three tillage systems in south central Honduras. Am J Altern Agric. 17:4–8.

Sebilo, S., J.M. Blair, and C.W. Edwards. 1998. Using anion exchange membranes to measure soil nitrate availability and net nitrification. Soil biol biochem. 27:914–917. doi:10.1016/S0038-0717(98)00020-3

Sullivan, D. 2005. Are "Heavy feeds" meaningful indicators of soil health, and estimators of nitrogen fertilizer credits. WERA-103 Conference Publication. Nutrient Digest V2-2.

Subler, J.P., J.D. Fernandez, S.A. Norton, E. Rustad, and A.L. White. 2006. Using ion-exchange resins to study soil response to experimental watershed acidification. Environ Monit Assess. 116:383–384. doi:10.1007/s10661-006-7112-3

Tabatabai, M.A., and J.M. Bremner. 1972. Arylsulfatase activity of soils. Soil Sci Soc Am J. 34:225–229.

To, X., D.L. Allan, C. Rosen, and D.T. Kaiser. 2015. Evaluation of the Soultz test as an indicator of potential nutrient releases in Minnesota soils. Proceedings of the Annual Meeting of the ASA-CSSA-SSSA, Minneapolis, MN. 15–18 Nov 2015. https://scisoc.confex.com/scisoc/2015am/webprogram/Paper95130.html (accessed 27 Jan 2016).

To, X., D.L. Allan, C. Rosen, and D.E. Kaiser. 2014. Soultz-b and other mineralizable nitrogen tests as indicators of N response in Minnesota soils. Presented at the Annual Meeting of the ASA-CSSA-SSSA, Long Beach, CA. 2–5 Nov 2014. https://scisoc.confex.com/scisoc/2014am/webprogram/Paper90276.html (accessed 27 Jan 2016).

Vadas, P.J., J.J. Spruochuta, and M. Sharp. 2006. Estimation of net N mineralization from plant-amended soil to predict N availability in N plant residue time-dependent. Soil Use Manage. 26:200–203. doi:10.1111/j.1475-2743.2006.00388.x

vander Salm, C.J., J. Dolfing, M. Heinen, and J.J. Schils. 2007. Estimation of nitrogen losses via denitrification from a heavy clay soil under grass. Agric Ecosyst Environ. 119:311–319. doi:10.1016/j.agee.2006.07.012

Wolf, R.R., K.K. Haney, M.A. Shiou, J.R. Deputy, and E.G. Beauchamp. 2000. Determining N-availability in manure with different methods of monitoring and indices. Soil Sci Soc Am J. Proc. 36:179–180. doi:10.2136/sssaj1972.03615995003600010044x

McLaughlin, M.J., P.A. Lancaster, P.W.G. Sale, N.C. Uren, and K.I. Peverill. 1993. Use of cation/anion exchange membranes for multi-element testing of acidic soils. Plant Soil 155–156:223–226. doi:10.1007/BF00025024

Mahony, D.C., T. Malzer, J. LaMorenaux, O.S. Walton, A.L. Gimenez, R. Mayer, and P.D. Woodruff, et al. 2006. In situ silicone tube microextraction: A new method for unobtrusive time-integrated sampling of redox biogeochemistry from undisturbed environments. In soil J. Chem. Geol. Sci. 276–1382. doi:10.1007/s10586-006-9718-6

Moron, A., and D. Cozzolino. 2003. Determination of potentially mineralizable nitrogen and nitrogen in particulate organic matter fractions in soil by visible and near-infrared reflectance spectroscopy. J Agric. Sci. 137:263–613. doi:10.1017/S0021859602002526

Murwanay, R.D., and S.A. Khan. 2001. Diffusion methods to determine different forms of nitrogen in soil hydrolysates. Soil Sci. Soc Am J. 65:1284–1292. doi:10.2136/sssaj2001.6541284x

Nelson, D.R. 1981. Microbial transformations of Amino Res. Biochem. 50:78–83. doi:10.1146/annurev.bi.50.070181.002555

Peoné, J.A., M.L. Carrera, and A.J. Franzleubbers. 2002. A rapid method to estimate potentially mineralizable nitrogen in soil. Soil Sci Soc Am J. 64:962–1847. doi:10.2136/sssaj2002.1847

Prokopowich, W.P., and R.H. Fox. 1992. Use of a cathode/PVC membrane to predict sidedress nitrogen. Agron J. 84:59–63. doi:10.2134/agronj1992.00021962008400010013x

Qian, P., and J.J. Schoenau. 2005. Use of ion exchange membrane to assess nitrogen-supply power of soils. J. Plant. Nutr. 28:2193–2200. doi:10.1080/01904160500323118

Ravim, K.S., M. Singh, A.K. Biswas, A. Swarup, and K.A. Dwivedi. 2001. Changes in organic and inorganic sulfur fractions and S-mineralization in a Typic Haplustert after long-term cropping with different fertilizer and organic manure inputs. Aust J Soil Res. 39:737–744. doi:10.1071/SR00020

Sahrawat, K.L., and N.N. Chittapuram. 2008. Exchange of ammonium, nitrate and others for enhancing the Total Extraction of T...

(Further references illegible.)

Selmer-Olsen, A.R., A.N. Oien, R. Baerug, and I. Lyngstad. 1971. The United States conversion of organic matter and nitrification rates with soil and sand. Soil Sci Soc Am J. 35:1–5. doi:10.2136/sssaj1971.03615995003500010011x

Sikora, L.J., and S.L. Smith. 1983. Nitrogen mineralization potentials of various soils. Soil Sci Soc Am. Proc. 36:465–472. doi:10.2136/sssaj1972.03615995003600030018x

Printed and bound by CPI Group (UK) Ltd, Croydon, CR0 4YY

27/10/2024

14580342-0001